Pilot Identification of Symbols and an Exploration of Symbol Design Issues for Electronic Displays of Aeronautical Charting Information

DOT/FAA/AR-07/37

DOT-VNTSC-FAA-07-07

Air Traffic Organization Operations Planning

Human Factors Research and Engineering Group

Washington, DC 20591

Divya C. Chandra

Michelle Yeh

U.S. Department of Transportation

Research and Innovative Technology Administration

John A. Volpe National Transportation Systems Center

Cambridge, MA 02142

June 2007

This document is available to the public through the National Technical Information Service, Springfield, Virginia, 22161.

Notice

This document is disseminated under the sponsorship of the Department of Transportation in the interest of information exchange. The United States Government assumes no liability for its contents or use thereof.

Notice

The United States Government does not endorse products or manufacturers. Trade or manufacturers' names appear herein solely because they are considered essential to the objective of this report.

REPORT DOCUMENTATION PAGE		*Form Approved*
		OMB No. 0704-0188

Public reporting burden for this collection of information is estimated to average 1 hour per response, including the time for reviewing instructions, searching existing data sources, gathering and maintaining the data needed, and completing and reviewing the collection of information. Send comments regarding this burden estimate or any other aspect of this collection of information, including suggestions for reducing this burden, to Washington Headquarters Services, Directorate for Information Operations and Reports, 1215 Jefferson Davis Highway, Suite 1204, Arlington, VA 22202-4302, and to the Office of Management and Budget, Paperwork Reduction Project (0704-0188), Washington, DC 20503.

1. AGENCY USE ONLY (Leave blank)	2. REPORT DATE June 2007	3. REPORT TYPE AND DATES COVERED
4. TITLE AND SUBTITLE Pilot Identification of Symbols and an Exploration of Symbol Design Issues for Electronic Displays of Aeronautical Charting Information		5. FUNDING NUMBERS FA6Y/DD305
6. AUTHOR(S) Divya C. Chandra and Michelle Yeh		
7. PERFORMING ORGANIZATION NAME(S) AND ADDRESS(ES) U.S. Department of Transportation John A. Volpe National Transportation Systems Center Research and Innovative Technology Administration Cambridge, MA 02142-1093		8. PERFORMING ORGANIZATION REPORT NUMBER DOT-VNTSC-FAA-07-07
9. SPONSORING/MONITORING AGENCY NAME(S) AND ADDRESS(ES) U.S. Department of Transportation Federal Aviation Administration Air Traffic Organization Operations Planning Human Factors Research and Engineering Group 800 Independence Avenue, SW Washington, D.C. 20591 Program Manager: Dr. Tom McCloy		10. SPONSORING/MONITORING AGENCY REPORT NUMBER DOT/FAA/AR-07/37
11. SUPPLEMENTARY NOTES		
12a. DISTRIBUTION/AVAILABILITY STATEMENT This document is available to the public through the National Technical Information Service, Springfield, VA 22161		12b. DISTRIBUTION CODE
13. ABSTRACT (Maximum 200 words) This report describes a study designed to address four research questions about symbology for electronic displays of charting information. The main research question was whether pilots could identify a set of test symbols. Three other research questions were addressed regarding: (1) labels for navigation aid symbols, (2) grouping navigation aids into families, and (3) line style conventions on paper charts and electronic moving map displays. The tested symbols are being considered for inclusion in an updated industry recommendations document, specifically, the SAE International Aerospace Recommended Practices (ARP) document on Electronic Aeronautical Symbols (ARP 5289). The Federal Aviation Administration or the International Civil Aviation Organization may choose to adopt this industry document by reference, at a later date. Note that this research applies to any electronic display that shows the symbology (i.e., symbols and lines) tested in this study, regardless of the intended function of the display.		
14. SUBJECT TERM Electronic displays, aeronautical charts, symbology, moving map displays, charting information, symbols		15. NUMBER OF PAGES 75
		16. PRICE CODE
17. SECURITY CLASSIFICATION OF REPORT Unclassified	18. SECURITY CLASSIFICATION OF THIS PAGE Unclassified	19. SECURITY CLASSIFICATION OF ABSTRACT Unclassified	20. LIMITATION OF ABSTRACT

This page left blank intentionally.

PREFACE

This report was prepared by the Human Factors Division of the Office of Aviation Programs at the Volpe National Transportation Systems Center. It was completed with funding from the FAA Human Factors Research and Engineering Group (AJP-61) in support of the Aircraft Certification Service Avionics Branch (AIR-130). We would like to thank our FAA program manager, Tom McCloy, as well as our technical sponsor Colleen Donovan. We would also like to thank all the members of the SAE International G-10 Aeronautical Charting Committee, who invested their expertise and time to make this study so valuable. Particular thanks go to the many pilots who donated their time and input for the study. In addition, we would like to thank Catherine Guthy, Andrea Sparko, and Hao Tran for assistance with the data processing, and Raquel Rodriguez, for assistance with distributing the surveys.

The views expressed herein are those of the authors and do not necessarily reflect the views of the Volpe National Transportation Systems Center, the Research and Innovative Technology Administration, or the United States Department of Transportation.

Feedback on this document may be sent to Divya Chandra (Divya.Chandra@volpe.dot.gov) or Michelle Yeh (Michelle.Yeh@volpe.dot.gov). Further information on this research effort can be found at http://www.volpe.dot.gov/hf.

METRIC/ENGLISH CONVERSION FACTORS

ENGLISH TO METRIC

LENGTH (APPROXIMATE)
1 inch (in) = 2.5 centimeters (cm)
1 foot (ft) = 30 centimeters (cm)
1 yard (yd) = 0.9 meter (m)
1 mile (mi) = 1.6 kilometers (km)

AREA (APPROXIMATE)
1 square inch (sq in, in^2) = 6.5 square centimeters (cm^2)
1 square foot (sq ft, ft^2) = 0.09 square meter (m^2)
1 square yard (sq yd, yd^2) = 0.8 square meter (m^2)
1 square mile (sq mi, mi^2) = 2.6 square kilometers (km^2)
1 acre = 0.4 hectare (he) = 4,000 square meters (m^2)

MASS - WEIGHT (APPROXIMATE)
1 ounce (oz) = 28 grams (gm)
1 pound (lb) = 0.45 kilogram (kg)
1 short ton = 2,000 pounds (lb) = 0.9 tonne (t)

VOLUME (APPROXIMATE)
1 teaspoon (tsp) = 5 milliliters (ml)
1 tablespoon (tbsp) = 15 milliliters (ml)
1 fluid ounce (fl oz) = 30 milliliters (ml)
1 cup (c) = 0.24 liter (l)
1 pint (pt) = 0.47 liter (l)
1 quart (qt) = 0.96 liter (l)
1 gallon (gal) = 3.8 liters (l)
1 cubic foot (cu ft, ft^3) = 0.03 cubic meter (m^3)
1 cubic yard (cu yd, yd^3) = 0.76 cubic meter (m^3)

TEMPERATURE (EXACT)
[(x-32)(5/9)] °F = y °C

METRIC TO ENGLISH

LENGTH (APPROXIMATE)
1 millimeter (mm) = 0.04 inch (in)
1 centimeter (cm) = 0.4 inch (in)
1 meter (m) = 3.3 feet (ft)
1 meter (m) = 1.1 yards (yd)
1 kilometer (km) = 0.6 mile (mi)

AREA (APPROXIMATE)
1 square centimeter (cm^2) = 0.16 square inch (sq in, in^2)
1 square meter (m^2) = 1.2 square yards (sq yd, yd^2)
1 square kilometer (km^2) = 0.4 square mile (sq mi, mi^2)
10,000 square meters (m^2) = 1 hectare (ha) = 2.5 acres

MASS - WEIGHT (APPROXIMATE)
1 gram (gm) = 0.036 ounce (oz)
1 kilogram (kg) = 2.2 pounds (lb)
1 tonne (t) = 1,000 kilograms (kg)
= 1.1 short tons

VOLUME (APPROXIMATE)
1 milliliter (ml) = 0.03 fluid ounce (fl oz)
1 liter (l) = 2.1 pints (pt)
1 liter (l) = 1.06 quarts (qt)
1 liter (l) = 0.26 gallon (gal)

1 cubic meter (m^3) = 36 cubic feet (cu ft, ft^3)
1 cubic meter (m^3) = 1.3 cubic yards (cu yd, yd^3)

TEMPERATURE (EXACT)
[(9/5) y + 32] °C = x °F

QUICK INCH - CENTIMETER LENGTH CONVERSION

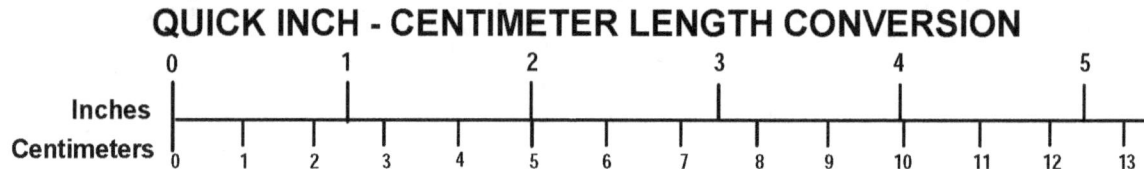

QUICK FAHRENHEIT - CELSIUS TEMPERATURE CONVERSION

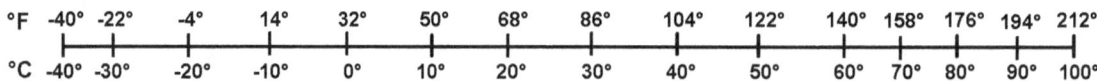

For more exact and or other conversion factors, see NIST Miscellaneous Publication 286, Units of Weights and Measures. Price $2.50
SD Catalog No. C13 10286   Updated 6/17/98

Table of Contents

List of Figures viii
List of Tables ix
Executive Summary xi
Acronyms xiii
1 Introduction 1
    1.1 Existing Guidance 2
    1.2 Goals for Industry Recommendations Document 2
    1.3 Symbol Stereotypes for Navigation Aids 3
2 Method 4
    2.1 Participants 4
    2.2 Procedure 4
3 Identification of Symbols 5
    3.1 Symbols Tested 5
    3.2 Task 8
    3.3 Analysis Process 9
    3.4 Results 10
    3.5 Summary of Results for Symbol Identification Task 16
4 Labels for Navigation Aid Symbols 19
    4.1 Task 19
    4.2 Results 20
    4.3 Statistical Analyses 21
    4.4 Summary and Discussion 22
5 Grouping Navigation Aids into Families 23
    5.1 Task 23
    5.2 Results and Discussion 23
6 Line Style Conventions 25
    6.1 Task 25
    6.2 Results and Discussion 26
7 Summary and Conclusions 33
8 References 34
Appendix A: Questionnaire A.1
Appendix B: Response Coding Method for the General-Symbol Identification Task ... B.1
Appendix C: Cluster Analysis Results C.1

List of Figures

Figure 1. Example of general symbol identification task. 9
Figure 2. Example of airport symbol identification task. 9
Figure 3. Example of the navigation-aid identification task. 19
Figure 4. Example of line-style question for a paper chart element. 26
Figure 5. Example of line-style question for a moving map display element. 26
Figure 6. Responses for how a radial or bearing line that defines an intersection is shown on paper charts. 27
Figure 7. Responses for how a procedure path is shown on paper charts. 28
Figure 8. Responses for how a transition path is shown on paper charts. 28
Figure 9. Responses for how airways are shown on paper charts. 28
Figure 10. Responses for how a missed approach path is shown on paper charts. 29
Figure 11. Responses for how visual flight paths are shown on paper charts. 29
Figure 12. Responses for how other conditional routes are shown on paper charts. 29
Figure 13. Responses for how airways are shown on electronic moving map displays. 30
Figure 14. Responses for how the active flight plan is shown on electronic moving map displays. 30
Figure 15. Responses for how an alternate flight plan is shown on electronic moving map displays. 31
Figure 16. Responses for how a missed approach path is shown on electronic moving map displays. 31
Figure 17. Responses for how airspace boundaries are shown on electronic moving map displays. 31
Figure 18. Responses for how a state or country boundary is shown on electronic moving map displays. 32

List of Tables

Table 1. Symbols included in the study. 6

Table 2. Original and simplified NDB shapes. 7

Table 3. NDB-LOM symbols. 7

Table 4. Obstructions symbols. 7

Table 5. Localizers. 8

Table 6. Other symbols. 8

Table 7. Airport symbols. 8

Table 8. Foil (i.e., fake) shapes included in the study 8

Table 9. Results for the NDB-LOM family symbol shapes. 10

Table 10. Results for obstruction symbol shapes. 11

Table 11. Reports of obstruction details. 12

Table 12. Results for localizer symbol shapes. 13

Table 13. Results for other symbols. 14

Table 14. Responses for symbols representing civil, military, or joint-use airports. 15

Table 15. Summary of results for symbol identification task for general symbols. 17

Table 16. Summary of results for airport symbols. 18

Table 17. Test conditions for navigation-aid identification task 19

Table 18. Mean accuracy of identification for navigation aids. 20

Table 19. Mean confidence score for navigation aids. 21

Table 20. Cluster analysis results. 23

Table 21. Features used to group navigation aid symbols and their relative frequencies. 24

Table 22. Linear display elements for which knowledge of line style conventions was assessed. 25

Table 23. Excerpt of data for low single obstruction symbol. B.2

This page left blank intentionally.

Executive Summary

This report describes a study designed to address four research questions about symbology for electronic displays of charting information. It is important to emphasize that this research applies to any electronic display that shows the symbology (i.e., symbols and lines) tested in this study, regardless of the intended function of the display.

The main research question was whether pilots could identify a set of test symbols selected by the SAE International Aerospace Behavior and Technology (SAE G-10) Aeronautical Charting Committee. The test symbols, which included 16 general symbols (e.g., obstructions and markers), and 6 airport symbols (e.g., heliport and seaport), are being considered for inclusion in an updated industry recommendations document, specifically, the SAE International Aerospace Recommended Practices (ARP) document on Electronic Aeronautical Symbols (ARP 5289). Most of the tested symbol shapes were identified as intended by a majority of participants. A few problematic symbols were noted, and factors contributing to their misinterpretation are discussed.

The second research question was whether labels increase the accuracy of identifying navigation aid symbols. To explore this topic, pilots' accuracy in identifying five navigation aid symbols was tested based on symbol shape alone, frequency information alone, or both shape and frequency information. For four of the five tested symbols, there was no difference in participants' accuracy of symbol identification with or without the labeling information. The results of this test are not definitive, and a more detailed study is necessary to understand the full effect of labeling information on both the speed and accuracy of symbol identification.

The third research question was about the similarities that pilots see between navigation aids. These similarities were explored in order to identify sets of symbols that could be represented by a single, generic, family symbol shape. Results showed that pilots grouped navigation aid symbols into categories primarily based on the type of information provided by the navigation aid.

The final research question was whether pilots recognize the line style conventions that are in use today on paper charts and electronic moving-map displays. Pilots were found to be fairly knowledgeable about line conventions on paper charts, where standards are well established. Responses to the questions about line conventions on electronic moving map displays were more varied, which could indicate either that pilots did not know the conventions, or that the conventions are not well established on these displays.

Results of this research are intended to be of use to the FAA, the International Civil Aviation Organization (ICAO), and other Civil Aviation Authorities. These organizations may choose to adopt, by reference, the symbology recommendations developed by the SAE G-10 Aeronautical Charting Committee at a later date. The results of this research are also intended to be of use to industry manufacturers who develop and/or depict symbology.

This page left blank intentionally.

Acronyms

ADF	Automatic Direction Finder
ANOVA	Analysis of Variance
ARP	Aerospace Recommended Practice
ATP	Air Transport Pilot
CAA	Civil Aviation Authority
DME	Distance Measuring Equipment
EFB	Electronic Flight Bag
EMD	Electronic Map Display
FAA	Federal Aviation Administration
FMS	Flight Management System
GPS	Global Positioning System
ICAO	International Civil Aviation Organization
IFR	Instrument Flight Rules
ILS	Instrument Landing System
IM	Inner Marker
LDA	Localizer-Type Directional Aid
LMM	Compass Locator at the Middle Marker
LOC	Localizer
LOM	Compass Locator at the Outer Marker
MM	Middle Marker
MSA	Minimum Safe Altitude
NACO	National Aeronautical Charting Office
NDB	Non-directional radio beacon
OM	Outer Marker
RTCA	Radio Technical Commission for Aeronautics
SDF	Simplified Directional Facility
SIDs	Standard Instrument Departure charts
STARs	Standard Instrument Arrival charts
TAA	Terminal Arrival Area
TACAN	Tactical Air Navigation, a military navigation facility
TSO	Technical Standard Order
US	United States
VFR	Visual Flight Rules
VOR	Very High Frequency (VHF) Omni-directional Radio Range
VORDME	A joint VOR and DME facility
VORTAC	A joint VOR and TACAN facility

This page left blank intentionally.

1 Introduction

This report describes a study designed to address four research questions about electronic symbols for aeronautical charting information. These symbols (e.g., navigation aids and instrument-approach symbols) are shown on many different flight-deck systems today. For example, moving map displays typically show navigation aid symbols and other aeronautical charting information, such as airspace boundaries. Electronic Flight Bags (EFBs) can also show charting information on electronic aeronautical charts, which are generally based on paper formats. (See Chandra, Yeh, Riley, & Mangold, 2003 for a discussion of electronic charts on EFBs.) Database-driven electronic aeronautical charts that are intended to *replace* paper charts are also in development. These sophisticated charts will provide customizable electronic access to *all* information shown on paper charts, including reference values that may only be used occasionally (e.g., minimum altitudes for various situations).

The electronic symbols used on these systems sometimes vary significantly. Although there are industry-recommended standards for symbols on electronic displays of aeronautical charting information (see Section 1.1 below), these standards are not always followed. In addition, some of the symbols that are in use currently are not well recognized by pilots (Yeh & Chandra, 2006).

The lack of consistency in chart and map symbology is not desirable from a human factors perspective. When different symbols are used to represent a single object or chart element, there is a risk of confusion. This is especially true because pilots could be using multiple sources of information at the same time (e.g., a moving map display *and* a paper or electronic chart), and these sources may use different symbology conventions. In addition, there is the risk of conflict when two manufacturers use the same symbol to represent different chart elements. Pilots could misinterpret a symbol if they rely upon knowledge of a different manufacturer's charts or moving map display. The variety of symbology in use also impacts the cost of pilot training in the airlines because airline operators need to ensure that pilots understand all of the symbology used. Even pilots who rent aircraft are impacted because they may be exposed to different displays and symbology with each rental.

Moving aeronautical charting symbology towards a higher level of standardization would help alleviate the problems noted above. In addition, standards could lower a manufacturer's development costs because the manufacturer would not have to design and test new symbols, and, if the standards are invoked or referenced by a regulatory document, the manufacturer could be comfortable that the proposed symbology would be found acceptable by their Civil Aviation Authority (CAA).

In order to support the development of more standardized symbology for electronic aeronautical charts and moving map displays, the SAE International Aerospace Behavioral Engineering Technology Committee (SAE G-10) Aeronautical Charting Committee, is updating Aerospace Recommended Practices (ARP) 5289, *Electronic Aeronautical Symbols* (SAE, 1997). This document contains recommendations for symbols that are primarily shown on charts used during operations under Instrument Flight Rules (IFR) (e.g., instrument approach plates, arrival and departure terminal charts, and enroute charts), although some of the symbols are also found on charts for use under Visual Flight Rules (VFR). The current ARP 5289 document also contains a table that illustrates the recommended symbol set alongside the symbol sets in use by several major manufacturers of paper charts and moving map displays. Line styles are also recommended in the document (e.g., for the missed approach track and airspace boundaries), and there are some general suggestions on using lines of different weights (heavy, medium, and light).

With funding and support from the Federal Aviation Administration (FAA), the John A. Volpe National Transportation Systems Center (Volpe Center) is working with the SAE G-10 Aeronautical Charting Committee in their efforts to update ARP 5289 (SAE, 1997). The Volpe Center's primary task is to determine whether pilots can correctly identify the committee's preliminary symbol proposals. In addition, other research issues related to electronic symbols for charting information are explored.

Note this research effort is independent of the type of electronic display on which the symbol is shown; the results are intended to apply regardless of the display's intended function. Therefore, the applicability of the results may be far reaching. Results of this research are intended to be of use to the FAA, the International Civil Aviation Organization (ICAO), and other CAAs. These organizations may choose to adopt, by reference, the symbology recommendations developed by the SAE G-10 Aeronautical Charting Committee. The results of this research are also intended to be of use to industry manufacturers who develop and/or depict symbology.

Research issues in the design of electronic symbology were documented in an earlier paper (Yeh and Chandra, 2004). Two earlier studies conducted as part of this research program were documented in Yeh and Chandra (2005). The most recent study is reported here. The research plans are coordinated with the FAA sponsors and the SAE G-10 Aeronautical Charting Committee. Real-world constraints, such as display technology and pilot knowledge are considered during experiment design and data collection.

Background for the current research study is presented below. First, existing design guidance for electronic symbology on displays is presented. Next, goals for the updated industry recommendations document are reviewed. Finally, one particularly relevant Volpe Center study on electronic navigation-aid symbols is reviewed (Yeh and Chandra, 2006).

1.1 Existing Guidance

Design guidance for electronic symbology is provided in RTCA (2003), SAE (1997), and ICAO (2001).

RTCA DO-257A (RTCA, 2003), which is invoked by the FAA's Technical Standard Order (TSO) C165 (FAA, 2003), provides requirements and recommendations for the design of electronic map displays (EMDs). This document includes some information on what symbology to show. Specifically, RTCA DO-257A states that:

The EMD shall display distinctive symbols for different fix types (waypoints, airports, VORs, NDBs, intersections) and the aircraft (ownship).

Notes:

1. These symbols make up the minimum required symbol set, as listed in Table 2-1.

2. If the input to the EMD does not distinguish between flight plan fix types (e.g., VOR vs. NDB), then the waypoint symbol is acceptable. However, if off-route fixes (e.g., VORs) are displayed, they must use the distinctive symbols appropriate for the fix type.

The other two documents mentioned above provide recommendations for what the symbol shapes should look like, if they are shown. The SAE International ARP 5289 (SAE, 1997) provides industry recommendations for electronic aeronautical chart symbology. The document contains recommendations for both symbols and lines. ICAO Annex 4 (ICAO, 2001) provides recommended symbology primarily for paper aeronautical charts, but it also includes recommendations for a few electronic symbols, based on the symbology in ARP 5289.

1.2 Goals for Industry Recommendations Document

The recommended symbols in the current ARP 5289 (SAE, 1997) were expected to be recognizable by qualified pilots. In addition, the symbols were thought to be simple shapes that could be drawn on the current display technology. Unfortunately, Yeh and Chandra (2006) found that pilots did not recognize some of the recommended symbols. Specifically pilots did not recognize the symbols recommended in ARP 5289 for the VOR, DME, TACAN, VORTAC, and VORDME. Also, in interviews with manufacturers of moving map displays, it became clear that some of the proposed symbols were difficult to draw on existing displays (e.g., the DME, VORDME, and NDB). Another reason for the lack of standardization may be that the display manufacturers were not aware of ARP 5289 because it was not invoked by an FAA or ICAO document. In any case, some of the symbols recommended in ARP 5289 are not in widespread use among manufacturers of electronic displays.

The SAE G-10 Aeronautical Charting Committee is now updating the recommended symbol set in ARP 5289 with the goal of recommending recognizable shapes that can be implemented on all displays. These proposals are being tested to ensure that they are recognizable, and input from display manufacturers is being sought to ensure that they can implement the symbols.

The revised ARP 5289 will provide the same type of information as in the 1997 version of the document (SAE, 1997). Specifically, recommendations will be made for several symbols that are used on various types of paper charts and electronic displays of charting information. Guidance will also be provided on line styles. This document will not address recommendations with regard to color of the symbology; all of the sample symbols in the document will be shown in black and white. This document also does not specifically address issues unique to three-dimensional displays. A table of symbols currently in use by several manufacturers will be provided in a separate document in preparation by the Volpe Center, an industry review of symbology.

1.3   *Symbol Stereotypes for Navigation Aids*

There are eight common navigation symbols shown on both aeronautical charts and moving map displays: DME, intersection/fix, NDB, TACAN, VOR, VOR/DME, VORTAC, and waypoint. Because of their importance, these eight symbols were the first research priority. A study was conducted to determine whether pilots recognized symbols for these eight navigation aids (see Yeh and Chandra, 2005 for full details; Yeh and Chandra, 2006 provides a summary). The Volpe Center gathered symbol shapes for these navigation aids, which were then in use, from several chart and display manufacturers. Shapes that were recommended for use by ICAO and SAE International (ICAO, 2001; SAE, 1997) were also included in the study.

The symbol shapes were placed into sets according to the type of navigation aid symbol. That is, all the DME symbol shapes were grouped, all the intersection/fix shapes were grouped, etc. Instrument-rated pilots, including many airline pilots, were asked to consider each group of symbol shapes separately. For example, the pilot might start with the set of DME symbol shapes. Within the set, he/she considered each shape one at a time, and decided whether it represented a DME or not (yes or no). Data from 73 pilots were obtained and aggregated. A statistical test was computed to determine whether a given symbol shape was considered "representative" of the set or not. In general, symbols that received a large number of *yes* responses were considered *representative* and symbols that received a large number of *no* responses were considered *unrepresentative*. Symbols with a mix of yes and no responses were neither representative nor unrepresentative. For details of the procedure and statistical test, see Yeh and Chandra (2005).

The results of this task identified well-recognized (stereotypical) symbol shapes for seven of the eight symbol types studied. The only symbol that did not have a highly recognized shape was the stand-alone DME. This was most likely because it is unusual to see a stand-alone DME. Most DMEs are co-located with another navigation aid, such as a VOR, and shown as a joint VORDME facility. Yeh and Chandra (2006) also found that some symbol shapes that are in use today were considered to be unrepresentative, largely because of their overall shape. Variations in color, fill, and other details generally did not impede pilots' ability to recognize the symbol.

The SAE G-10 Aeronautical Charting Committee plans to include the navigation-aid symbol shapes that were found to be representative by Yeh and Chandra (2006) in the revised ARP 5289 document, with one minor modification to the NDB symbol shape. Some of the dots in the tested NDB shape were deleted in order to make it easier to depict on electronic displays. The revised symbol was retested in the study reported here. In the case of the DME symbol, where no shape was statistically determined to be representative, the committee plans to recommend a square, which was one of the most familiar shapes for a stand-alone DME.

2 Method

Having tested eight common navigation aid symbols in detail in the previous study (Yeh and Chandra, 2006), the next research priority was to determine whether pilots could identify other symbols proposed by the industry committee for commonly seen elements (e.g., obstructions, markers, localizers, as well as airports). Three other research questions were also addressed in this study:

1) <u>Labels for navigation aid symbols</u>. Do labels increase the accuracy of identifying navigation aids?

2) <u>Grouping navigation aids into families</u>. What similarities do pilots see between navigation aids? How would pilots group the symbols into "families," where a family could be represented by a single "generic" symbol on the electronic display?

3) <u>Line style conventions on paper charts and electronic moving map displays</u>. Do pilots recognize line style conventions that are in use today on paper charts and electronic moving map displays?

Paper-and-pencil tasks were developed to address each of the four research questions above and incorporated into a four-part survey (see Appendix A). The tasks for each of these questions are described in detail later in the report. Participants in the study and the general procedure for the study are described below.

2.1 Participants

Ninety-six instrument rated pilots (88 male, 6 female, and 2 who did not submit gender information) participated in the study. Almost all of the participants (84) held an Air Transport Pilot (ATP) rating. Their flight experience ranged from 800 to 25,000 flight hours, with an average of 8300 flight hours. The participants had a variety of flight experience, though the majority (62) were active airline pilots. Seventeen pilots had military flight experience, and 37 had either private business or corporate flight experience.[1] Many of the pilots had a mix of flight experience. For example, some airline pilots also had military experience.

Seventy-six pilots indicated that they primarily used charts from Jeppesen and 20 indicated they primarily used charts from the United States Government National Aeronautical Charting Office (NACO). Ninety-one pilots indicated experience with glass cockpits, 90 with Flight Management Systems (FMS) displays, and 82 with moving map displays.

2.2 Procedure

Pilots were recruited through the FAA, the Air Line Pilots Association (ALPA), the military (specifically, the Air Force Advanced Instrument School), and a local general aviation flying club. The study was advertised in a short verbal or written announcement (e.g., at a meeting or in a newsletter), and interested pilots were asked to contact the Volpe Center to have a paper questionnaire mailed to them. Participants did not receive any compensation for their participation.

---

[1] The intention was to distinguish between private flights conducted for business purposes ("business") from flights conducted by a corporate flight department ("corporate"). However, it is not known whether this distinction was clear to participants.

Overall, 125 questionnaires were returned. Unfortunately, 29 of these had to be discarded, in most cases because the consent forms were not signed, leaving 96 useable responses.[2] The bulk of the data came from responses to an electronic newsletter posting to ALPA members, which explains the high number of airline pilot participants. Only a few questionnaires were obtained from purely general aviation pilots, for whom the study was not broadly advertised. The overall response rate was not tracked.[3]

3 Identification of Symbols

The goal of this task was to determine whether pilots could correctly identify the symbols proposed by the SAE G-10 Aeronautical Charting Committee. The committee constructed the test symbol set based on symbols in use today. The proposed symbols may not be familiar to all pilots because they were taken from a variety of sources. The group intends to use the results of this study to help develop a final set of recommended symbols. For background information on the symbols and how they are used in flight operations, see two FAA reference documents (FAA 2007a, 2007b).

3.1 Symbols Tested

The symbols were divided into two sections in the survey. The first section included 18 general symbols, 16 of which were real and two of which were fake symbols, called foils. The second section included eight airport symbols, six of which were real and two of which were foils. The foils were not real symbols, but they were *realistic* symbols. Responses to the foil symbols provide insights into how familiar symbol features might influence the recognition of unfamiliar shapes.

A list of the symbols tested is given in Table 1. One shape was tested for most of the symbol types. However, two shapes were tested for three particular symbols, the generic localizer, MSA, and NDB. In these cases, two shapes were tested in order to provide the SAE G-10 Aeronautical Charting Committee with more information about how pilots interpreted the symbols, as explained further below.

---

[2] Pilots who did not sign the consent forms presumably thought they were ensuring their anonymity. However, the signed consent form, which assures that pilot that his/her participation is strictly confidential, is what gives us permission to use the data. The Institutional Review Board who reviewed the study procedure to ensure that participants are treated ethically requires the signed form. In addition, anonymity of the data was assured because the consent forms were separated from the rest of the data upon receipt.

[3] The overall response rate was difficult to track because the materials for the study were distributed freely at different venues via different methods. For example, the materials were provided to anyone who requested them at various meetings, regardless of whether the recipient was a qualified pilot. In some cases, a single individual at an organization was allowed to photocopy the materials, resulting in an unknown number of copies being distributed. In another case, the materials were left out on a desk for pilots to pick up, and it is not known how many of the surveys were actually taken.

General Symbols	**Airport Symbols**
Airport Beacon	Civil airport
Back Course localizer	Closed airport
Generic Localizer Candidates, 2 shapes tested	Heliport
Holding Pattern	Joint civil-military airport
Locator Outer Marker (LOM)	Military airport
Marker (same shape for inner, middle, and outer marker)	Seaport
Minimum Safe Altitude (MSA), 2 shapes tested	
Multiple High Obstruction	
Multiple Low Obstruction	
Non-directional Beacon (NDB), 2 shapes tested	
Single High Obstruction	
Single Low Obstruction	
Terminal Arrival Area (TAA)	

Table 1. Symbols included in the study.

3.1.1 Selection Criteria

As explained earlier, shapes for seven common navigation aids (the DME, intersection/fix, TACAN, VOR, VOR/DME, VORTAC, and waypoint) were tested earlier and agreed upon based on that study (Yeh and Chandra, 2006), so they were excluded from the current study.

Additional criteria for symbol selection are explained below. The test symbol shapes are shown in the next section.

- The NDB symbol was included in this study because it was suspected that the original shape identified as stereotypical for an NDB in Yeh & Chandra (2006), shown on the left in Table 2, could not be depicted easily on electronic displays. Therefore, a simplified version of the shape, shown in Table 2, was constructed and tested here. The SAE G-10 Aeronautical Charting Committee was interested to see whether results of the earlier study, that the NDB shape with the circular array of dots was more easily recognized, would be replicated with the less dense array of dots.

- The two candidates for generic localizer symbols (i.e., shapes that represent either a front or back course localizer) were based on variants of similar symbols in use today, which are shaped like long narrow arrowheads. One of the generic localizer shapes tested is currently used by NACO to represent a Simplified Directional Facility (SDF) or Localizer-Type Directional Aid (LDA) approach, which are both uncommon procedures. The SAE G-10 Aeronautical Charting Committee was not sure whether that symbol was well known. If not, it could potentially be reassigned to indicate a generic localizer symbol. The second localizer shape had a similar outline as the first one, but included more detail.

- Two MSA shapes were tested in order to obtain data to discriminate between reasonable alternatives: the current ICAO symbol, and the current NACO symbol.

- Symbols whose shapes were so generic that they were not expected to be identifiable in isolation were generally excluded (e.g., a plain circle for an airport).

- The joint military-civil-use heliport symbols was excluded from the study because it occurs rarely and was expected to be unfamiliar to most pilots, meaning that recognition frequency would not be informative. There is no current symbol in use for a "military heliport," although one of the two airport-foil symbols incorporated features of both a military airport (the double ring), and a civil heliport (the "H" in the center of the symbol).

- The SAE G-10 Aeronautical Charting Committee did not have proposed symbols for use on the profile view of instrument approach charts ready for testing, so they were not considered.

Original	Simplified shape
(dotted circle with center dot)	(dotted circle)

Table 2. Original and simplified NDB shapes.

3.1.2 Test Symbol Shapes

The general symbol shapes selected for the study can be grouped into four categories: NDB-LOM symbols, obstructions, localizers, and other symbols. These groups are described below, and illustrated in the accompanying tables. The airport symbols and foil shapes are also shown below.

NDB-LOM Symbols (Table 3). Three symbol types from the NDB-LOM family were tested: LOM, marker, and NDB. A simplified NDB shape, NDB (1) was included, as discussed above. In addition, a second NDB shape, NDB (2) was tested. NDB (2) is a double ring that is used on some electronic displays.

Obstructions (Table 4). Four shapes representing obstructions were tested. The four shapes distinguish between the height of the obstruction (high vs. low), and whether the obstruction is a single object or group of objects. Note that not all chart providers currently distinguish between low and high obstructions via the symbol shape.

Localizers (Table 5). Three shapes representing localizers were tested, one for the back course localizer, and two candidates for a generic localizer symbol, as discussed above.

Other Symbols (Table 6). Five other shapes representing a beacon, holding pattern, MSA, and TAA were tested. Two shapes for an MSA were tested. The first shape, labeled MSA (1), is recommended by ICAO, and the second shape, labeled MSA (2), is used in FAA (NACO) charts.

Airport Symbols (Table 7). The six airport symbol shapes tested are shown in Table 7.

Foil Shapes (Table 8). The foil shapes are shown in Table 8. As noted above, these shapes are not currently in use, but they do incorporate features of real symbols.

LOM	Marker	NDB (1)	NDB (2)
(symbol)	(symbol)	(symbol)	(symbol)

Table 3. NDB-LOM symbols.

Low, single obstruction	Low, group obstructions	High, single obstruction	High, group obstructions
(symbol)	(symbol)	(symbol)	(symbol)

Table 4. Obstructions symbols.

7

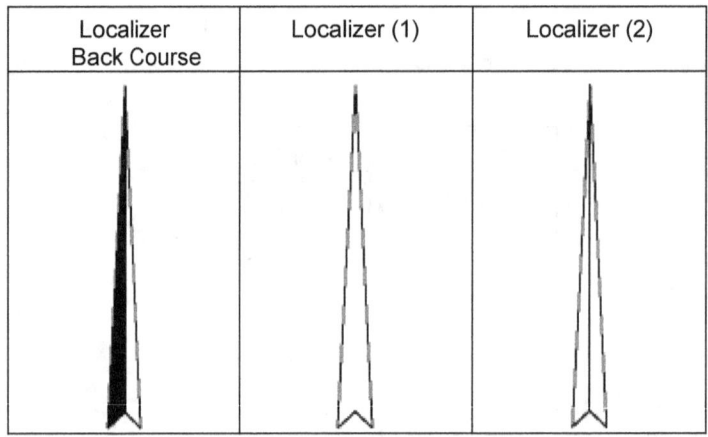

Table 5. Localizers.

Table 6. Other symbols.

Table 7. Airport symbols.

Table 8. Foil (i.e., fake) shapes included in the study.

3.2 Task

The test was conducted in the form of a paper-and-pencil survey. Pilots were shown each symbol shape without any context, and asked to identify and write (i.e., recall) what the shape represented, and how confident they were in their response. Specifically, the instructions read:

> *For each shape below, identify it and indicate your level of confidence in your response. Some of the symbols are unusual, so you may not recognize all the symbols. Write "?" if you do not know or can't tell.*

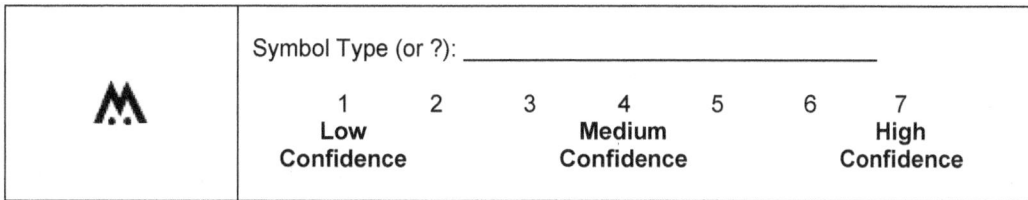

Figure 1. Example of general symbol identification task.

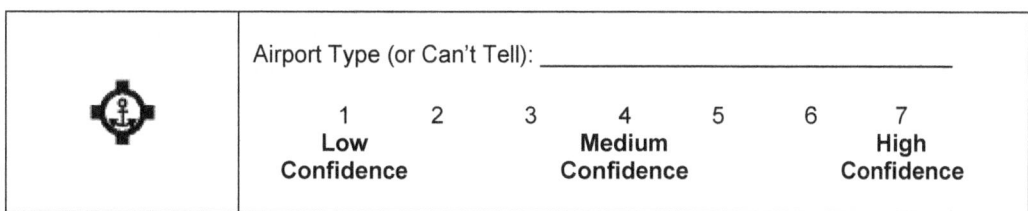

Figure 2. Example of airport symbol identification task.

The second sentence was intended to reduce the chance that pilots would be frustrated by the inclusion of symbols that they did not recognize (i.e., the fake shapes, called foils).

Two examples are shown below in Figure 1 and Figure 2. For the general symbols, the expected response was a name/description for the symbol type (e.g., *group of obstacles*). For the airport symbols, participants were asked to identify the *type* of airport. In other words, participants were told that the symbol represented an airport, and they were asked to provide more details about the kind of airport that was depicted (e.g., "seaport").

3.3 Analysis Process

Responses for the symbol identification task varied because participants could write in any response. In order to make sense of this variety, response *categories* were constructed for each symbol, based on the range of individual responses found for that symbol. For the general symbols, each individual response was then sorted into *one* of the response categories. For details on the process of coding the data for the general symbols into response categories, see Appendix B. For the airport symbols, response categories were framed in terms of properties mentioned about the airport, and participants' responses were sorted according to the attributes mentioned. Some responses mentioned more than one attribute, in which case the response was counted towards both attributes. For example, a response such as *controlled airport with services* was counted towards both the *Air Traffic Control* category and the *Services* categories.

Tables showing the frequency of responses in each category were then constructed. These tables illustrate the range of responses but, by themselves, do not indicate whether the responses were "correct" or "acceptable." In order to assess whether the responses were acceptable, a judgment call was made by the subject-matter-experts in the SAE G-10 Aeronautical Charting Committee. For example, for the obstruction symbol, the *Manmade Structure* response category was judged to be an acceptable response, but the *Terrain* response was not, because terrain is represented differently (e.g., by shaded contours). In some cases, however, acceptability of a response was not clear-cut. There were some responses that were understandable, but not technically correct. These situations are discussed further below, as they arise.

Results of the analysis indicate which symbols were well recognized and which were confusing. Symbols that were well recognized had a high proportion of responses in the correct response category, and their average confidence ratings were high relative to those for the confusing symbols. Symbols that were confusing either had responses that were spread out over several different response categories, or had a substantial proportion of responses in an incorrect category. The confusing symbols also typically had a high rate of *Can't Tell* and/or missing responses.

3.4 Results

The holding pattern and the two MSA shapes were recognized by all or nearly all of the participants with high confidence. The holding pattern and MSA (1), the ICAO version in Table 6, were both identified correctly by 100% of the pilots. The average confidence rating for the holding pattern was 6.67 out of 7, and 6.6 out of 7 for MSA (1). MSA (2), the NACO version in Table 6, was identified correctly by 97% of participants and had an average confidence rating of 6.37. Responses to the two MSA shapes differed slightly in that more of the responses to MSA (2) mentioned detail about its center point (26% versus 14% for MSA (1)).

Three other general shapes were well recognized: the Marker, NDB (1), and the LOM. Detailed results for these shapes are given in Table 9. The Marker shape was identified correctly by 90% of pilots, with an average confidence of 6.0 out of 7. In coding responses to the Marker symbol, distinctions between the inner marker, middle marker, and outer marker were ignored, because the same symbol shape is planned to be recommended for all of these elements. NDB (1), the shape with the circular array of dots, was identified as an NDB by 83% of pilots, with an average confidence of 6.01. This finding confirms that the simplified shape for the NDB symbol tested here is sufficiently similar to the more detailed shape identified as stereotypical for the NDB in Yeh and Chandra (2006). The LOM shape was identified correctly by 70% of pilots with an average confidence of 6.17. There were no missing responses for these three shapes.

Symbol	Mean Confidence Rating (Max 7)	Response Categories	Frequency (% of Responses)
Marker	6.0	A. Marker, Marker Beacon, MM, OM, IM, fan marker, beacon alone	**90%**
		B. Other	4%
		Both A and B	3%
		Can't Tell	3%
NDB (1)	6.01	NDB/ADF[4], compass locator, beacon	**83%**
		Other	4%
		Can't Tell	13%
LOM	6.17	LOM, LMM, NDB and marker, compass locator with marker	**70%**
		Compass locator alone, beacon alone	13%
		Marker alone	10%
		NDB alone	1%
		Other	1%
		Can't Tell	5%

Table 9. Results for the NDB-LOM family symbol shapes.

---

[4] ADF (Automatic Direction Finder) is the name of the instrument on the aircraft that displays the location of the NDB relative to aircraft position, and is often mistakenly used to refer to the NDB instead.

Among the airport symbols, almost all participants correctly identified the heliport, closed-airport, and seaport shapes. Specifically, the heliport shape was identified by 98% of participants, with an average confidence of 6.31 out of 7; the shape representing a closed airport was identified by 95% of participants with an average confidence of 6.0; and the shape representing a seaport was identified by 94% of pilots with an average confidence of 5.93.

The remaining symbols, which received more varied responses, are discussed in the next several sections.

3.4.1 Obstruction Shapes

Responses to symbol shapes representing obstructions are shown in Table 10. The results indicate that participants were generally able to identify these symbol shapes as either *Obstructions* or *Manmade Structures* and they did so with a fairly high degree of confidence.

The SAE G-10 Aeronautical Charting Committee judged that either of these response categories (*Obstruction* or *Manmade Structure*) was acceptable. However, the committee did not intend for these symbols to represent a natural obstacle, such as terrain, and that response category was considered to be unacceptable. A small but significant number of *Terrain* responses to the two "low" symbols were therefore an indicator that those symbol shapes are potentially misleading. Neither of the two "high" symbols yielded any *Terrain* responses. These results are likely because of the potential inference that the low-obstruction symbol resembles a mountain peak, whereas the high-obstruction symbol resembles a manmade tower.

Responses to the obstruction symbol shapes were also examined to determine the number of pilots who correctly reported *detail* about the meaning of the symbol. Specifically, detail about the height (low versus high) and multiplicity (single versus group) of the obstruction was examined. Table 11 below shows that, regardless of the shape of the symbol, correct detail about the number of obstructions (single versus group) was reported more often than correct detail about the height of the obstructions (74% versus 34% of the time). This implies that the height of the obstruction was not conveyed by the symbol shape as well as information about the number of objects.

Response Categories	Frequency (% of Responses)			
	Low Single Obstruction	Low Group Obstruction	High Single Obstruction	High Group Obstruction
A. Obstruction	52%	46%	29%	33%
B. Manmade Structures	30%	29%	62%	56%
C. Terrain	8%	13%	--	--
Both A and B	3%	4%	4%	1%
Both A and C	2%	5%	--	--
Both B and C	1%	1%	1%	1%
Other	1%	--	--	--
Missing	2%	2%	3%	8%
Overall Frequency of Acceptable Responses (A, B, or Both A & B)	85%	79%	95%	90%
Mean Confidence Rating (Max 7)	5.96	6.08	6.20	6.19

Table 10. Results for obstruction symbol shapes.

Response Details	Overall Frequency	Frequency (% of Responses with the detail)			
		Low Single Obstruction ⋀	Low Group Obstruction ⋀⋀	High Single Obstruction	High Group Obstruction
Reports of Number Detail (single versus group)	70%	69%	63%	71%	78%
Reports of Height Detail (low versus high)	34%	26%	22%	49%	38%

Table 11. Reports of obstruction details.

3.4.2 Localizer Shapes

Responses to the localizer shapes are shown in Table 12 below. A majority of participants (65%) correctly identified the symbol shape representing a back-course localizer as a *Localizer*. Almost half of these correct responses (31% of the total responses) included correct detail, that the shape represented a *back course*. The second most common response for the back-course localizer symbol shape was *ILS*, given by 28% of pilots. The ILS response was considered to be incorrect by the SAE G-10 Aeronautical Charting Committee because the real ILS shape, which indicates availability of vertical guidance, is completely shaded or screened, not just shaded or screened on one side, as was the symbol tested in this study. However, the ILS symbol and the localizer symbol shapes do have a common outline. In addition, instrument approach charts with ILS procedures show a relatively large localizer symbol on the plan view (i.e., birds-eye view) of the chart, and they show a relatively smaller ILS symbol in the profile view of the chart, which depicts altitudes for the descent. Although incorrect, it is understandable that some pilots interpreted the localizer symbol to be an ILS symbol because of the prominence of the localizer shape on the instrument approach chart that is labeled as an ILS procedure. Note also that participants indicated a high confidence in their responses to the back-course localizer symbol shape.

Responses to the Localizer (1) and Localizer (2) shapes are more varied than responses to the back-course localizer shape. As mentioned earlier, these two shapes are candidates for the generic localizer shape. Localizer (1) is the shape currently used as an *SDF/LDA* shape on NACO charts, and 24% of pilots correctly identified it as such. This is an unexpectedly high rate of recognition for such an uncommonly used symbol and procedure. Just 22% of pilots responded in the way that the SAE G-10 Aeronautical Charting Committee expected, calling Localizer (1) a *Localizer*. An even larger proportion, 36% of pilots, said they could not tell what the shape represented. Given the number of *Can't Tell* and SDF/LDA responses, this symbol is confusing at best.

Responses to the Localizer (2) shape leaned more favorably towards the expected *Localizer* response (42%). Fewer pilots confused this shape with an *SDF/LDA* (10%), or with an *ILS* (10%). In addition, there was a moderate rate of *Can't Tell* (17%); this was better than the 36% rate for Localizer (1), but worse than the 1% rate obtained for the back-course localizer shape.

Response Categories	Localizer Back Course	Localizer (1)	Localizer (2)
A. Localizer	65%	22%	42%
B. SDF/LDA	--	24%	10%
C. ILS	28%	3%	10%
D. Other	2%	7%	9%
E. Can't Tell	1%	36%	17%
Both A and B	--	1%	4%
Both A and C	4%	1%	3%
Both B and D	--	1%	--
A, B, C, and D	--	--	1%
Missing	--	4%	3%
Mean Confidence Rating (Max 7)	**6.43**	**4.52**	**4.91**

Table 12. Results for localizer symbol shapes.

There is also another, more positive, way to interpret the identification rates for the two generic localizer candidates. The rates of identification, while low relative to other more familiar symbols, could be seen as relatively good, considering that the symbols are new to pilots. This point of view assumes that pilots are highly attuned to all the details in a symbol, and the fact that they were able to extrapolate their knowledge to these test symbols was not necessarily expected. In other words, *despite* their novelty, the familiar outlines of the two localizer-symbol shapes did help some pilots to infer their meaning as expected by the industry committee.

3.4.3 Confusing General Symbols

Responses for the NDB (2), TAA, and Beacon symbol shapes varied. Results for these shapes are shown in Table 13.

The most common response for NDB (2) was *Can't Tell* (42%). The shape was identified as either a military or generic airport by 33% of pilots. In fact, the shape is very similar to that of a military airport (also a double circle, but with a less space between the two circles), and somewhat similar to the generic airport symbol (a single circle) that is used by some display manufacturers. The instructions did not say that airport symbols were excluded, so, without any other context, the airport response is not unreasonable. Although some transport display manufacturers do use this shape to represent an NDB, it was identified as an NDB by just 8% of pilots. NDBs are rarely used in commercial operations, so the airline pilots in this study probably had little, if any, experience with this shape in that context. With an average rating of 4.17, pilot confidence was low for NDB (2), confirming its unfamiliarity.

The most common response to the TAA shape shown in Table 13 was *MSA* (58%). The symbol shape for a TAA was described correctly by just 14% of participants. TAAs may not be as familiar to pilots as MSAs because they are used only on area navigation (RNAV) approaches. The MSA is visually similar to a TAA symbol (see Table 6), but an MSA is, in fact, different from a TAA (see FAA, 2007a and 2007b). Both the TAA and MSA provide obstacle clearance, but in addition, a TAA provides altitude information for the transition from the enroute structure to the terminal area with minimal air traffic control communications.

The most common response to the star shape shown in Table 13 was *Can't Tell* (47%). The star shape was intended to represent an airport beacon, but was described correctly as such by only 30% of participants. A few participants made responses that could be associated with the symbol shape (control tower, or facility not in continuous operation). Often, a text label is shown next to the star shape to provide context. Without the symbol or label, the beacon symbol may have lost much of its meaning.

Symbol	Mean Confidence Rating (Max 7)	Response Categories	Frequency (% of Responses)
NDB (2)	4.17	NDB/ADF, compass locator, beacon	8%
		Military Airport	14%
		Airport (generic)	19%
		Other	15%
		Can't Tell	42%
		Missing	3%
TAA	5.47	TAA/Terminal Arrival Area/Terminal Arrival Altitude	14%
		MSA	58%
		Other	14%
		Can't Tell	10%
		Missing	4%
Beacon	4.98	A. Airport Beacon	30%
		B. Control tower	4%
		C. Facility not in continuous operation	7%
		D. Other	8%
		E. Can't Tell	47%
		A and B	1%
		A, B, and C	1%
		Missing	1%

Table 13. Results for other symbols.

3.4.4 Confusing Airport Symbols

There was variability amongst the responses to the symbol shapes for civil, military, and joint civil-military airports. Responses for these symbols are given in Table 14. Recall that, for airports, a single response might have been counted towards more than one category, so the sum of the percentages in each column can exceed 100%.

As Table 14 shows, less than half the participants identified the distinction between the proposed shapes for military (42%), civil (30%), and joint civil-military (23%) airports. While it may be argued that the distinction between civil and military airports is important primarily to military pilots, examination of the data from pilots with military experience only showed no better identification of these symbols. Participants' confidence in their responses to these three airport symbols was generally neutral.

It is likely that these differences are not often pertinent to flight operations, particularly for the majority of the participants who were airline pilots. As a result, pilots may not have much experience distinguishing between these civil, military, and joint-use airports.

It is interesting to note that the feature used by NACO charts to represent services such as fuel at the airfield (i.e., squares at the 0°, 90°, 180°, and 270° points) were relatively well identified. Detail regarding *services*, *facilities*, or *fuel* was included in the responses of 32% of pilots for the civil airport symbol, 25% for the joint civil-military airport symbol, and 19% for the seaport symbol.

Response Category	Civil	Military	Joint civil-military
Correct Airport Type	30%	42%	23%
Services	32%	4%	25%
Runway parameter (length or surface)	17%	6%	9%
Air Traffic Control	5%	5%	6%
IFR/VFR	3%	3%	2%
Other (including incorrect airport type)	9%	11%	22%
Can't Tell	18%	31%	34%
Missing	2%	1%	4%
Mean Confidence Rating (Max 7)	5.37	5.06	4.78

Table 14. Responses for symbols representing civil, military, or joint-use airports.

3.4.5 Foils

The starburst-shaped foil in shown in Table 8 was not identified by 86% of participants; 75% of pilots said *Can't Tell* and another 11% provided no response. Those few who did provide a response indicated that the shape represented some type of hazard area (5%), some special kind of waypoint (5%), or some type of lighted object (3%). There is some logic to these guesses, based on the physical shape of the object, but clearly it is not a compelling fake symbol.

Thirty-seven percent of participants did not identify the three-pointed star foil shape in Table 8 (32% said *Can't Tell* and 5% provided no response), but 53% of the participants did call it a waypoint. In a previous study (Yeh & Chandra, 2006) the stereotypical shape for a waypoint was identified as a four-pointed star;

a three-pointed star, which was included as a foil in that study was not considered to be representative of a waypoint based on a statistical test.

The airport foil that included components of a military airport symbol and civil heliport symbol in Table 8 was determined to be a heliport of some sort by 59% of participants. Of these, almost half (44%) said that the symbol represented a military heliport. Thus, a significant number of respondents (26% of the total sample) used the two familiar features of the symbols to create a logical response, even though such an airport is actually fictional; no unique symbol for military heliport is in use. Note also that 33% of participants gave a *Can't Tell* response, and 2% did not answer the question. While these rates may seem high, they are, in fact, similar to the Can't Tell and missing rates for the real civil, military, and joint civil-military airport symbols shown in Table 14.

The last foil shape, a circle with two crossing lines in it also received some interesting responses. The most common responses focused on the two crossing lines, which had the appearance of runways. Runway properties such as configuration, length, or surface type were mentioned in 43% of responses. The second most common response (20%) referred to the user of the airfield (e.g., civil/military, private, or glider airport). One-quarter (25%) of responses were *Can't Tell*, and 2% of responses were missing for this foil shape.

3.5 Summary of Results for Symbol Identification Task

The results for the general symbol shapes are summarized in Table 15 below, which shows the most well recognized symbols towards the top, and the most confusing symbols towards the bottom. Results for the airport symbol identification task are summarized in a similar manner in Table 16. Symbols that were recognized 70% of the time or better just show the percent of correct responses. Peaks in the response distribution are presented for symbols that were less well recognized. The last five symbols in Table 15 are considered to be problematic relative to the others in the test because their responses are more varied. These shapes, discussed in detail above, are: the two candidates for a generic localizer, Localizer (1) and Localizer (2), the airport beacon, the TAA, and one of the two NDB shapes, NDB (2).

Name of Symbol	Symbol Shape	Peaks in Response Distribution	Mean Confidence Rating (Max 7)	Name of Symbol	Symbol Shape	Peaks in Response Distribution	Mean Confidence Rating (Max 7)
Holding Pattern		100% correct	6.67	Localizer (2)		42% correct 10% SDF/LDA 10% ILS 17% Can't Tell	4.91
MSA (1) (ICAO version)		100% correct	6.6				
MSA (2) (NACO version)		97% correct	6.37	Airport Beacon		30% correct 47% Can't Tell	4.98
Multiple High Obstruction		95% correct	6.19	Localizer (1) (NACO symbol for an SDF/LDA)		24% SDF/LDA 22% Localizer 36% Can't Tell	4.52
Marker		90% correct	6.00				
Single High Obstruction		90% correct	6.20				
Single Low Obstruction		85% correct	5.96				
NDB (1)		83% correct	6.01	TAA		10% correct 58% MSA 14% Can't Tell	5.47
Multiple Low Obstruction		79% correct	6.08	NDB (2)		42% Can't Tell 33% Military or general airport	4.17
LOM		70% correct	6.17				
Back Course Localizer		65% correct 28% ILS	6.43				

Table 15. Summary of results for symbol identification task for general symbols.

Name of Symbol	Symbol Shape	Peaks in Response Distribution	Mean Confidence Rating (Max 7)
Heliport	(H)	98% correct	6.31
Closed Airport	⊗	95% correct	6.0
Seaport	⚓	94% correct	5.93
Civil	○	30% correct 32% reference to services 18% Can't Tell 16% runway parameter (length or surface)	5.37
Military	◎	42% correct 31% Can't Tell	5.06
Joint Civil-Military	●	23% correct 34% Can't Tell 25% reference to services 22% other or incorrect airport type	4.78

Table 16. Summary of results for airport symbols.

4 Labels for Navigation Aid Symbols

4.1 Task

On paper charts, navigation aid symbols are often accompanied by supporting text labels, showing information such as the name of the facility, a shortened identifier (ID), the Morse code for the letters of the identifier, etc. However, if all this information is shown next to the symbol on an electronic chart, the resulting display is likely to be cluttered. Therefore, the research question here is what help does the text label provide in identifying the navigation aid.

In this experiment, only a small portion of the text label was studied. Specifically, stimuli for this task (i.e., the images that were presented) either did or did not show the frequency of the navigation facility, which provides a clue as to the type of facility.[5] A DME, which provides distance information, uses UHF frequencies; these frequencies are written as a *channel* (e.g., "Chan 31"). A VOR, which provides bearing information, uses a VHF frequency, similar to a broadcast commercial radio station (e.g., "103.8"). A facility that provides both distance and bearing information, such as a VORTAC, would show both a channel, and a VHF frequency.

Participants were asked to identify the navigation aid based on the information shown. An example question is shown in Figure 3, which shows a VORTAC symbol. Participants were asked to include as much detail as possible in their answers, and in particular to try to distinguish the symbol types (e.g., between VORs and VORDMEs). The accuracy of identifying the navigation aid (but not the speed of identification) was recorded, along with participants' ratings of their confidence in their response on a scale from 1 (low confidence) to 7 (high confidence).

Five navigation aids were tested: DME, TACAN, VOR, VORDME, and VORTAC. The test stimuli consisted of either the symbol shape alone, the frequency alone, or the symbol shape and frequency together, as shown in Table 17 below.

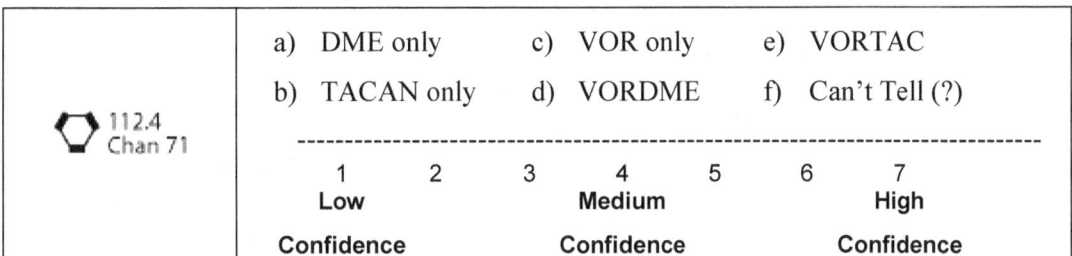

Figure 3. Example of the navigation-aid identification task.

Test Condition	Example
Symbol Shape alone	⬡
Frequency alone	113.8
Symbol Shape + Frequency	⬡ 112.6

Table 17. Test conditions for navigation-aid identification task.

---

[5] When the questionnaire was designed, the plan was to consider the symbol identifier as well as the frequency. These trials were included in the survey but they were deleted from the final analysis after consultation with the SAE G-10 Aeronautical Charting committee because it was determined that the symbol identifier did not provide a reliable clue about the facility type.

4.2 Results

The mean accuracy for each navigation aid is presented in Table 18 for each level of information.

Symbol Type	Symbol Alone	Frequency Alone	Symbol + Frequency	Mean Accuracy
DME	48%	Chan 31 80%	Chan 26 45%	57%
TACAN	65%	Chan 31 80%	Chan 88 76%	73%
VOR	81%	113.8 79%	112.6 83%	80%
VORDME	71%	113.8 115.8 Chan 105 82%	108.2 Chan 19 63%	71%
VORTAC	87%	113.8 115.8 Chan 105 82%	112.4 Chan 71 88%	85%

Table 18. Mean accuracy of identification for navigation aids.

Participants' confidence in their responses was evaluated using a *confidence score* to discriminate between participants who were both confident and correct from those who were confident but incorrect. The mean confidence score is presented in Table 19.

The confidence score was calculated by multiplying the participants' actual confidence *rating* (+1 to +7) by +1 if the answer was correct, and by −1 if the answer was incorrect. In other words, all correct responses yield positive confidence scores, and all incorrect responses yield negative confidence scores. The confidence score also distinguishes between participants who had low confidence in an incorrect answer and those who had high confidence in an incorrect answer. Participants who had low confidence in their incorrect answer yielded a *higher* confidence score than participants who had high confidence in their incorrect answer. For example, an incorrect response with a low confidence rating of 2 would have a confidence score of −2, but an incorrect response with a high confidence rating of 7 would have a lower confidence score of −7.

Symbol Type	Symbol Alone	Frequency Alone	Symbol + Frequency	Mean Confidence Score
DME	☐ 1.88	Chan 31 4.07	☐ Chan 26 0.83	2.26
TACAN	⬠ 2.63	Chan 31 4.07	⬠ Chan 88 3.52	3.41
VOR	⬡ 4.44	113.8 3.77	⬡ 112.6 4.53	4.25
VORDME	⬨ 4.20	115.8 113.8 Chan 105 2.45	⬨ 108.2 Chan 19 2.66	3.10
VORTAC	⬡ 5.50	115.8 113.8 Chan 105 2.45	⬡ 112.4 Chan 71 5.03	3.96

Table 19. Mean confidence score for navigation aids.

4.3 Statistical Analyses

Statistical analyses were performed on participants' accuracy and confidence scores to compare the three levels of *information provided* (symbol shape alone, frequency alone, and symbol plus frequency). In addition, the data were analyzed with respect to pilot experience with a particular *chart provider* because participants' familiarity with the symbols may vary depending on the charts used.

An Analysis of Variance (ANOVA) was performed on the accuracy and confidence scores for each navigation aid.[6] The analysis showed no statistical difference in accuracy as a function of the level of information for TACANs, VORs, VORDMEs, or VORTACs ($p > 0.05$). However, there was a statistically significant difference in accuracy for identifying DMEs depending on the level of information provided [$F(2, 279) = 7.98$, $p < 0.05$]. Paired comparison tests showed that the DME was classified most accurately when frequency information was presented alone ($p < 0.05$). This finding may be because the DME symbol shape by itself was not as familiar to participants, a theory that is accordance with the results from Yeh and Chandra (2006). In that study, some pilots commented that they never used the DME alone.

For confidence scores, there were no statistically significant differences attributable to the level of information for DMEs, TACANs, or VORs ($p > 0.05$). Therefore, even though the confidence scores for these symbols shapes may appear to vary in magnitude, from a statistical point of view the scores are all equivalent. However, there was a statistically significant difference in confidence score attributable to the level of information provided for VORDMEs [$F(2, 251) = 3.72$, $p < 0.05$] and VORTACs [$F(2, 275) = 7.77$, $p < 0.05$]. For both of these navigation aids, paired comparison tests showed that participants' confidence score was lowest when only frequency information was shown ($p < 0.05$). This finding is most likely because these two navigation aids cannot be distinguished from each other by frequency information alone. As shown in Table 19 above, the test items for the VORDME and VORTAC with

---

[6] This was a 3x2 ANOVA with 3 levels of information (symbol alone, frequency alone, symbol + frequency) and 2 levels of chart provider (Jeppesen, NACO).

frequency information alone are identical. For these test items, both answers were accepted, but participants' confidence in their response suffered.

The chart provider that the pilots used generally did not affect their responses for DMEs, TACANs, VORs, or VORTACs ($p > 0.05$). However, there was a statistically significant difference for VORDMEs: participants familiar with NACO charts were more accurate than those familiar with Jeppesen charts (85% vs. 68%, respectively) [$F(1, 279) = 7.85, p < 0.05$] and they were more confident as well, as shown by a higher confidence score (4.29 vs. 2.73, respectively) [$F(1, 251) = 3.72, p < 0.05$]. This finding is most likely a statistical anomaly due to the relatively low number of NACO users in the current study.[7] To confirm this interpretation, data from this study were compared with data from a larger sample of NACO users collected in an earlier study (Yeh and Chandra, 2006), in which participants were asked to identify the representative shape for each navigation aid. Although accuracy was not measured in that study, the VORDME symbol shape presented in the current questionnaire was considered to be representative of a VORDME by 82% of NACO users and 64% of Jeppesen users; this difference was *not* statistically significant. Therefore, the statistical difference for the identification of the VORDME symbol shape in this study is believed to be an anomaly.

4.4 Summary and Discussion

With the exception of the DME symbol, the tested navigation aid symbol shapes could be correctly identified based on shape alone. This is consistent with the finding from earlier Volpe Center research (Yeh and Chandra, 2006), that pilots are not especially familiar with the symbol for a stand-alone DME, and that the other navigation-aid symbol shapes tested are well recognized.

The tested navigation aid symbols could also be identified correctly based on frequency alone. The DME symbol was actually classified most accurately when only frequency information was shown. However, for two of the symbols, the VORDME and VORTAC, pilots' confidence in their response was lower when they were only shown the frequency information alone, without the symbol shape, most likely because the two are indistinguishable from one another based on frequency information alone.

The results suggest that combining shape information with frequency information does not significantly improve the identification accuracies above the rates achieved with either shape alone or frequency information alone. However, it is important to note that these results are far from definitive. Only a small set of symbols were tested; labels may be more meaningful for other symbols that were not tested. In addition, the rates of accurate identification may not be sensitive enough to detect differences in performance. Examination of response time, i.e., how long participants took to identify the test items, may provide more detail on the value of labels. Finally, this test only examined the use of frequency information. Other information, such as the name or identifier, especially if it is recognized, may be more or less helpful.

---

[7] Only 20 of the 96 participants reported that they primarily used NACO charts.

5 Grouping Navigation Aids into Families

5.1 Task

The research question addressed here was to understand what similarities pilots perceived among navigation aids. If similarities can be identified, symbol families can be created such that each family could be represented by a single "generic" symbol. Some manufacturers already use generic symbols for groups of navigation aid symbols (e.g., a VORDME, VORTAC, and VOR are all represented by the same shape), but the rationale for how to create the groups is not known. The goal here was to gather data on the relationships that pilots see between the navigation aids, so that these can be considered when grouping the symbols into families.

Participants were presented with the eight primary navigation symbols (DME, fix, NDB, TACAN, VOR, VORDME, VORTAC, and waypoint). They were asked to divide these symbols into groups of two, three, four, and five. Participants were also asked to give a title for each group to help explain their rationale for dividing the symbols. The groups and titles were used to identify relationships between the navigation aids.

5.2 Results and Discussion

Pilots' groupings of navigation aids were represented using similarity matrices that show the number of times two navigation aids were grouped together. (The matrices are provided and explained in Appendix C.) The relationships between the items were then analyzed using a cluster analysis to determine the common groupings. The results of the cluster analysis are shown in Table 20 below for each level of classification (two, three, four, or five groups). Dendrograms that depict the strength of the relationship among navigation aids are provided in Appendix C.

Two Groups

Group 1	Group 2
Fix, Waypoint	DME, NDB, TACAN, VOR, VORDME, VORTAC

Three Groups

Group 1	Group 2	Group 3
Fix, Waypoint	DME, NDB, VOR	TACAN, VORDME, VORTAC

Four Groups

Group 1	Group 2	Group 3	Group 4
Fix, Waypoint	NDB, VOR	DME	TACAN, VORDME, VORTAC

Five Groups

Group 1	Group 2	Group 3	Group 4	Group 5
Fix, Waypoint	NDB	DME	VOR	TACAN, VORDME, VORTAC

Table 20. Cluster analysis results.

As Table 20 shows, two groups were consistent throughout the categorization task. One group consisted of the fix and the waypoint, and the other consisted of the TACAN, VORDME, and VORTAC. The DME, NDB, and VOR did not have consistent groups. The NDB tended to be grouped with the VOR, but as the number of categories participants were asked to form increased, the NDB and VOR were separated into their own groups.

The titles that participants gave for their groups were subjectively coded in terms of the underlying features that were used to categorize the navigation aids. Seven features were used for this analysis, as listed and defined in Table 21 below. The overall number of times each feature was used by the subjects was tallied. As an example, when asked for two groups of symbols, one subject used the titles *Point in Space* and *Azimuth and/or Distance*. These titles were judged to be given on the basis of what information the navigation aid provided. The titles became more complex when there were more groups to name. For example, one subject used the following titles for the four-group condition: *Azimuth only, Azimuth and/or Distance, Military, Waypoint*. These titles were based on three features: *what information* was provided, *how much information* was provided, and who was the *intended user*. When the participant formed two or three groups, up to two features were coded for each title. When four or five groups were formed, up to three features of the titles were coded.

The results of the feature tally are shown in the four right-most columns of Table 21, which shows that the feature that participants considered most often when grouping the navigation aids was *what information* the navigation aid provided. This feature was used most often regardless of the number of navigation aid groups. (To see how the use of features varied by the number of navigation aid groups, read across the rows of Table 21.)

Feature	Definition	Number of Navigation Aid Groups			
		2	3	4	5
What Information	What type of information is provided? (e.g., distance information, radial/bearing information, or the latitude/longitude of a point)	58	59	56	52
How much Information	Is only one type of information provided, or more than one? Are both radial/bearing information and distance provided, or just one of these?	7	28	32	30
Utility	What is the utility of the information provided? (e.g., How much precision is there?, How often do you use it?, Is it easy to use? Is it only used in particular conditions?)	7	13	9	12
Technology	How is the information generated (by what technology, age of technology)?	16	18	23	16
Intended user	Who uses the information (e.g., civil/military, GA vs. transport)?	2	6	13	18
Unable to code	The titles were based on the names of the item (e.g., NDB, VOR-Type, etc). There was not enough information to establish what factor was used to separate the categories.	6	4	4	3
Missing/Can't Tell	Includes both *Can't Tell* (?) and no entry.	9	11	13	19
	Column Total	**105**	**139**	**150**	**150**

Table 21. **Features used to group navigation aid symbols and their relative frequencies.**

Table 21 also shows that as the number of groups increased, the total number of features participants considered also increased. This illustrates that as the number of groups increased, participants used more sophisticated rules to construct the groups. For example, as the number of groups participants were asked to form increased, more participants used features such as how much information was provided by the technology, the technology used, and the intended user (e.g., TACANs are used primarily by the military). These trends are seen by reading down the columns of Table 21.

6 Line Style Conventions

6.1 Task

The goal here was to examine whether pilots have knowledge of line style conventions used on paper charts and electronic moving map displays. Table 22 shows a list of the linear display elements that were assessed. For each element, pilots were asked to choose what type of line was used to depict that linear display element.

A sample question for a paper chart linear display element is provided in Figure 4 below, and a sample question for an electronic moving map display linear display element is shown in Figure 5. As shown in the figures, examples of the line styles were provided for items relevant to paper charts, but these were just schematic, not exact representations of the lines used on paper charts. The SAE G-10 Aeronautical Charting Committee agreed the sample line styles for paper charts were helpful even though they were not specific to a particular manufacturer. No examples were provided for moving map displays. It was expected that there would be more variability among line styles on moving map displays, and no representative styles for these lines were known. In addition, the response choices for the moving map displays did not distinguish between short, medium, and long dashed lines; only one dashed-line choice was given. All of the questions for paper charts had the same set of responses, and all of the moving map display questions had the same set of responses.

Paper Charts	**Moving Map Displays**
Radial or bearing line that defines an intersection	Missed approach path
Procedure path	Airways
Missed approach path	Alternate (non-activated) flight plan
Transition path	State or country boundary
Visual flight path	Active flight plan
Other conditional routes	Airspace boundary
Airways	

Table 22. Linear display elements for which knowledge of line style conventions was assessed.

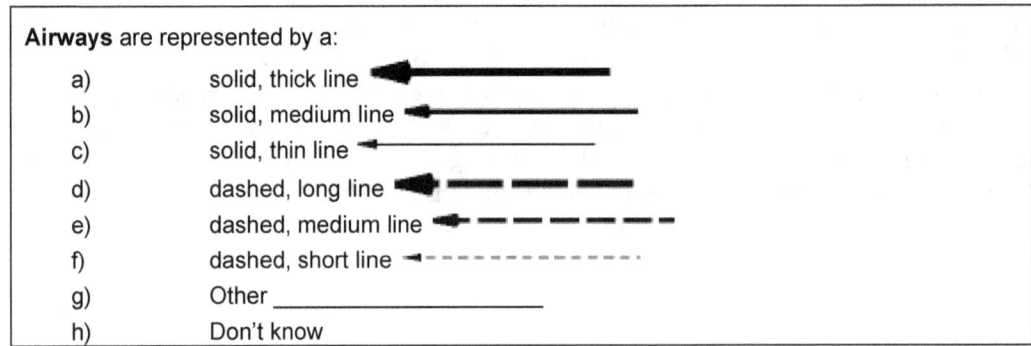

Figure 4. Example of line-style question for a paper chart element.

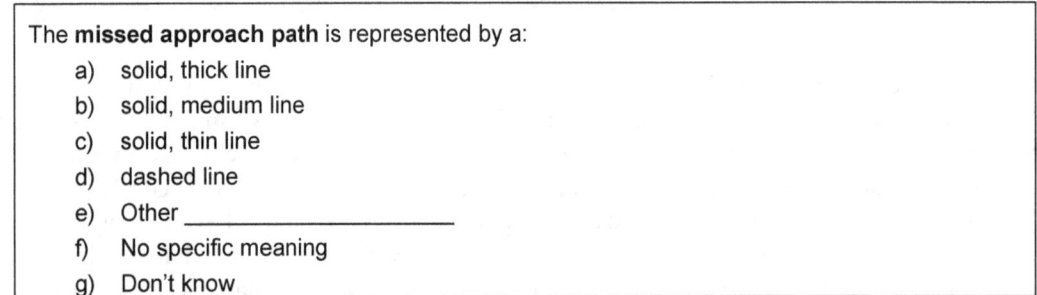

Figure 5. Example of line-style question for a moving map display element.

Most of the linear display elements tested for paper charts are familiar to pilots, but some are less familiar than others. In particular, *visual flight paths* and *other conditional routes* are relatively uncommon. A visual flight path might be seen on an instrument approach into a mountainous region. The visual flight path indicates that the runway must be in sight to proceed with the approach. Another type of visual flight path may occur when a route is denoted in terms of visual landmarks on the ground (e.g., "left turn at the fuel tanks"). The term *other conditional routes* represents a variety of uncommonly used paths. For example, a route to follow when the aircraft has lost communication with air traffic control is a conditional route. (On NACO charts, a lost communications path is represented as a dotted line, but other conditional routes are not depicted.) Jeppesen may depict conditional routes that are temporary routes set up to accommodate large public events, or temporary routes set up to accommodate navigation aids that are out of service.

The linear display elements tested for moving map displays were selected without extensive formal knowledge of which were commonly depicted. Although participants reported what moving map display they were most familiar with, there is currently no inventory that indicates what symbols all the different avionics vendors and map displays actually show, and how they are depicted. (As mentioned earlier, an effort to inventory current moving map display symbology is planned.)

6.2 Results and Discussion

Pilots' knowledge and perception of the line conventions and stereotypes for paper and electronic moving map displays are best depicted through histograms, which illustrate the response trends.

6.2.1 Paper charts

The histograms of participants' responses to line styles for paper chart linear display elements are presented in Figures 6 through 12 below. The most important feature to note in these graphs is the tightness of the response distribution. Histograms that show strong peaks, even though they may be across more than one response, indicate generally good consensus. If the peak covers two responses, it is most likely because participants could not clearly distinguish between the two responses. For example, participants did not always distinguish between the different thicknesses of the lines or the lengths of the dashes. For solid lines, participants distinguished between thick and thin lines but did not consistently distinguish between *thick* versus *medium* lines or between *medium* versus *thin* lines. For dashed lines, participants did not consistently distinguish between *medium* and *short* dash lengths.

The majority of participants considered a *radial or a bearing line that defines an intersection* (Figure 6) to be drawn with a solid, thin line (83%) and a *procedure path* (Figure 7) to be drawn with a solid, thick line (72%). These two responses are correct, in that they are in agreement with current paper chart depictions; NACO and Jeppesen do in fact use solid thin lines for a radial or bearing line that defines and intersection, and solid thick lines for procedure paths.

For other paths, responses were more spread out. Participants' responses showed that they considered a *transition path* to be drawn with a solid, medium or thin, line (Figure 8). NACO uses a solid medium line to depict a transition path, and Jeppesen uses a dashed medium line. *Airways* (Figure 9) were considered to be drawn with a solid, medium or thick, line. Both NACO and Jeppesen currently use a solid medium line to depict airways.

Participants' responses indicated that they considered a *missed approach path* (Figure 10) to be drawn with a dashed, medium or short, line. NACO uses a dashed short line. Jeppesen uses a thick dashed line (response *G*, or possibly *D*). The *visual flight path* (Figure 11) was thought to be drawn with dashed lines, with an approximately equal number of participants considering the dashes to be long, medium, or short. NACO uses a dashed line that could be interpreted as either medium or long for visual flight paths. Jeppesen uses a series of short bold arrows (response *G*).

The results showed no consensus for line conventions for *conditional routes* (Figure 12), a result that is not surprising when considering how broad a category this description represents. In addition, NACO does not use a single line style to represent this type of route, so pilots who were only familiar with NACO charts could not be expected to pick a single best answer. Jeppesen charts depict other conditional routes with a dashed medium line (response *E*).

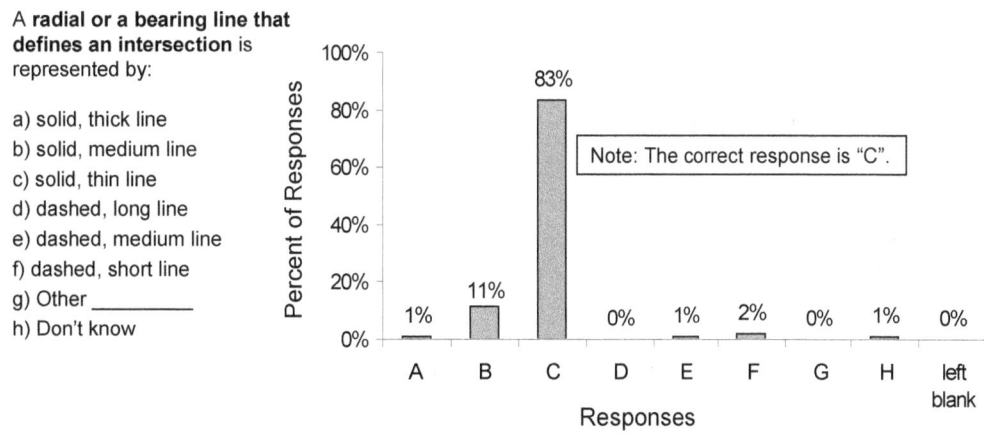

Figure 6. Responses for how a radial or bearing line that defines an intersection is shown on paper charts.

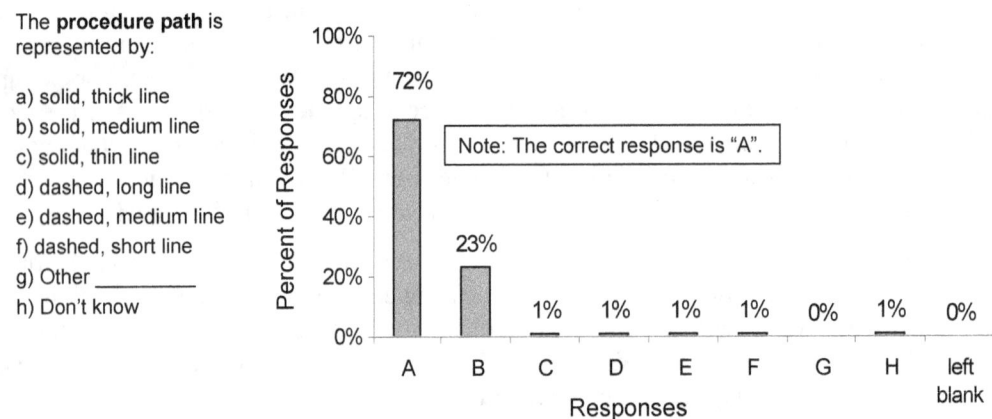

Figure 7. Responses for how a procedure path is shown on paper charts.

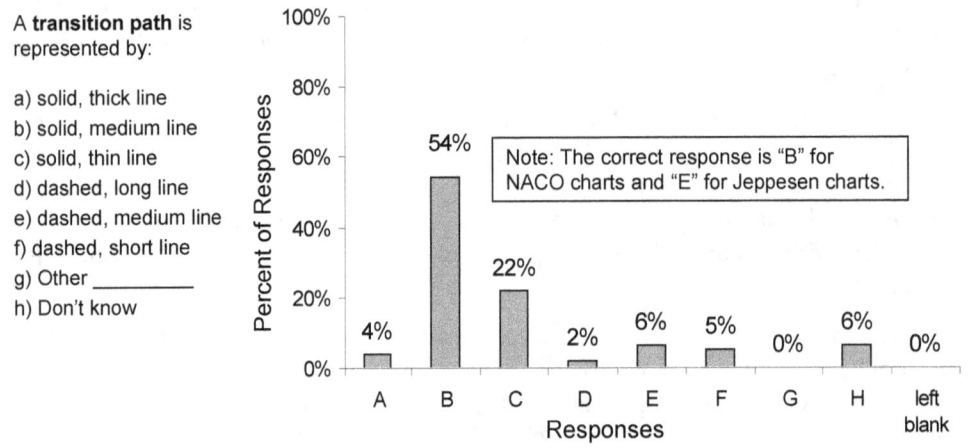

Figure 8. Responses for how a transition path is shown on paper charts.

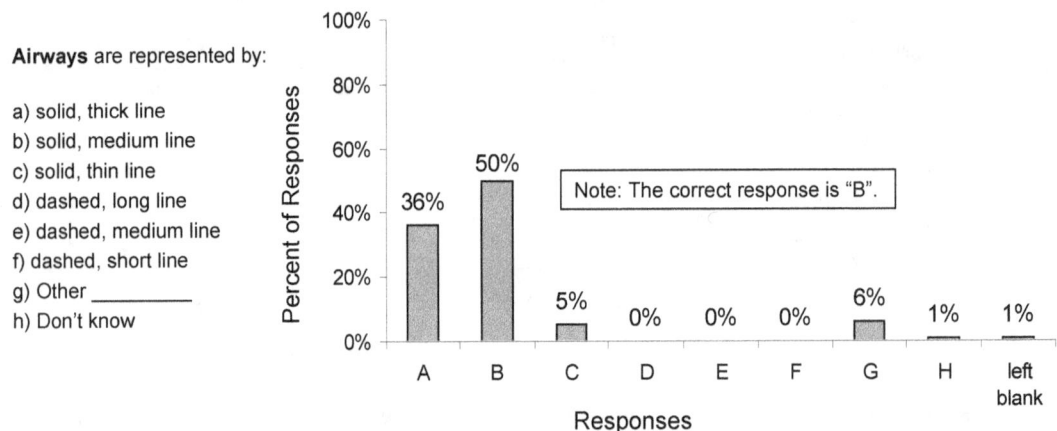

Figure 9. Responses for how airways are shown on paper charts.

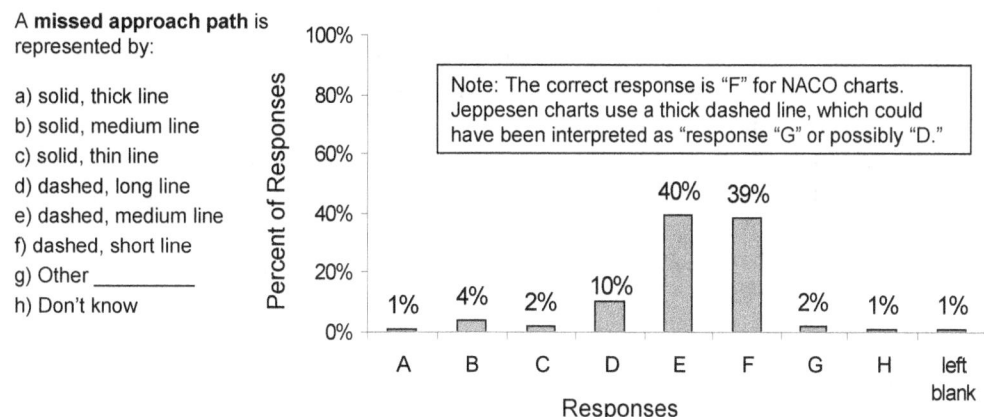

Figure 10. Responses for how a missed approach path is shown on paper charts.

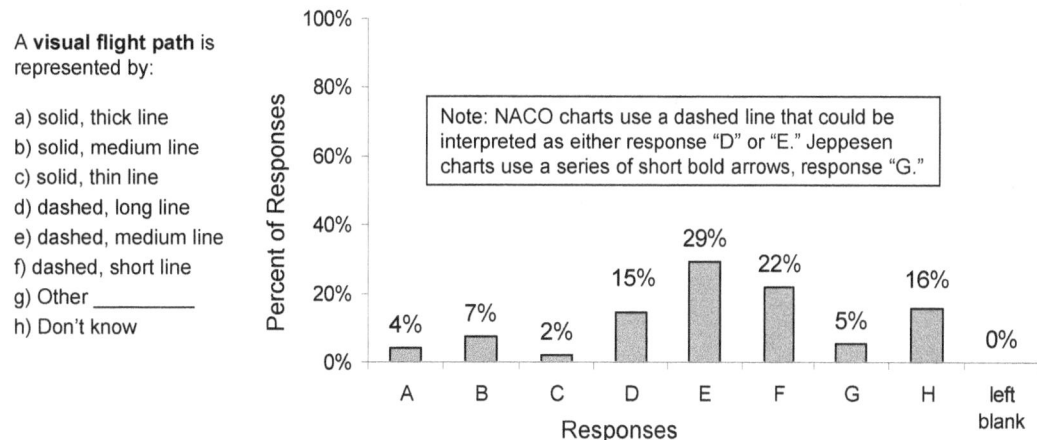

Figure 11. Responses for how visual flight paths are shown on paper charts.

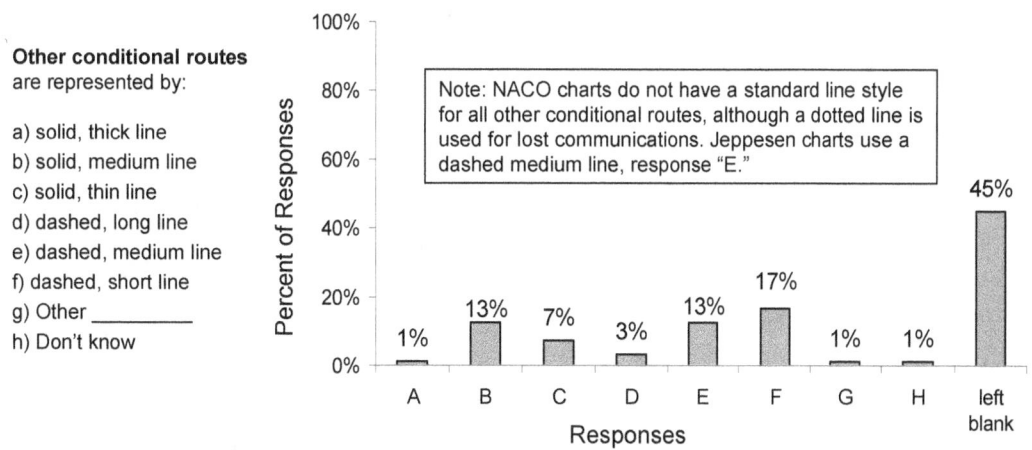

Figure 12. Responses for how other conditional routes are shown on paper charts.

6.2.2   Electronic Moving Map Displays

Participants' responses to line style conventions used on electronic moving map displays are shown in Figures 13 through 18 below, which show mixed responses to the questions. The larger variation in responses may be because there are more manufacturers of moving map displays than paper charts, and the symbology across these displays varies more than the symbology on various paper charts does. Other possibilities may be that pilots do not remember the moving map display line styles as well as the paper lines styles, that maps may not show all of these types of line styles, or that the electronic maps use other features, such as color, rather than thickness and dashing of the line style to distinguish lines. It was not possible to reconcile the responses with specific line styles used in actual electronic moving maps displays.

Even with the larger response variability, there was general consensus that *airways* (Figure 13) are drawn with a solid line (61%), although participants did not distinguish between the line thicknesses. In addition, the *active flight plan* (Figure 14) was generally thought to be a solid line of thick to medium thickness (63% of participants). There was no consensus for the conventions used for representing the *alternate (non-activated) flight plan* (Figure 15), the *missed approach path* (Figure 16), *airspace boundaries* (Figure 17), or *state or county boundary* (Figure 18).

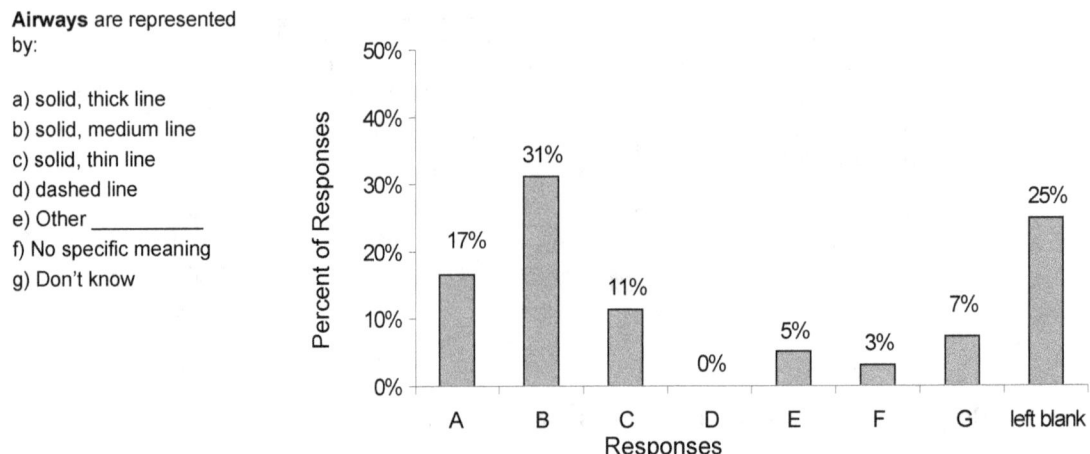

Figure 13. Responses for how airways are shown on electronic moving map displays.

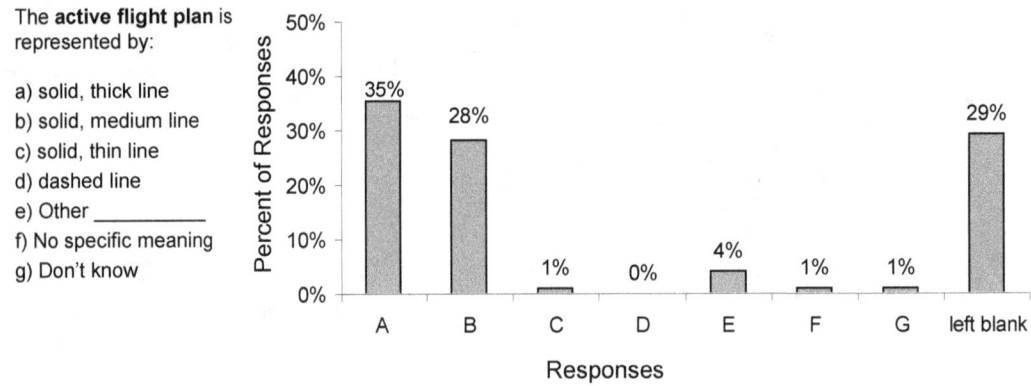

Figure 14. Responses for how the active flight plan is shown on electronic moving map displays.

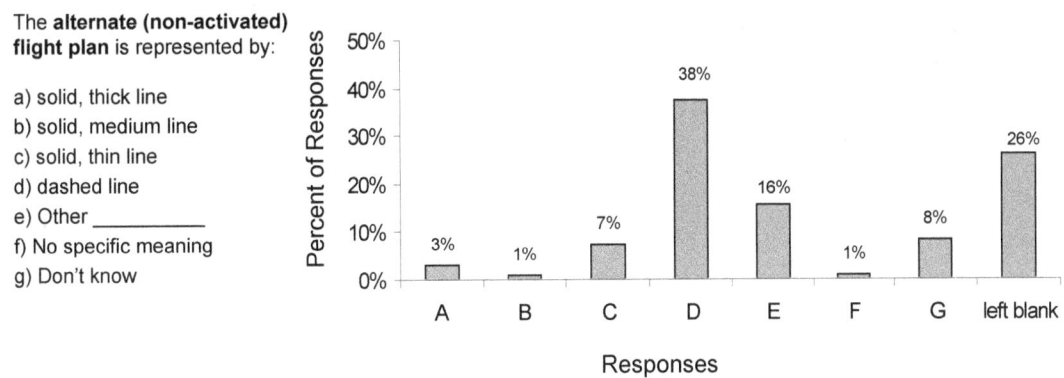

Figure 15. Responses for how an alternate flight plan is shown on electronic moving map displays.

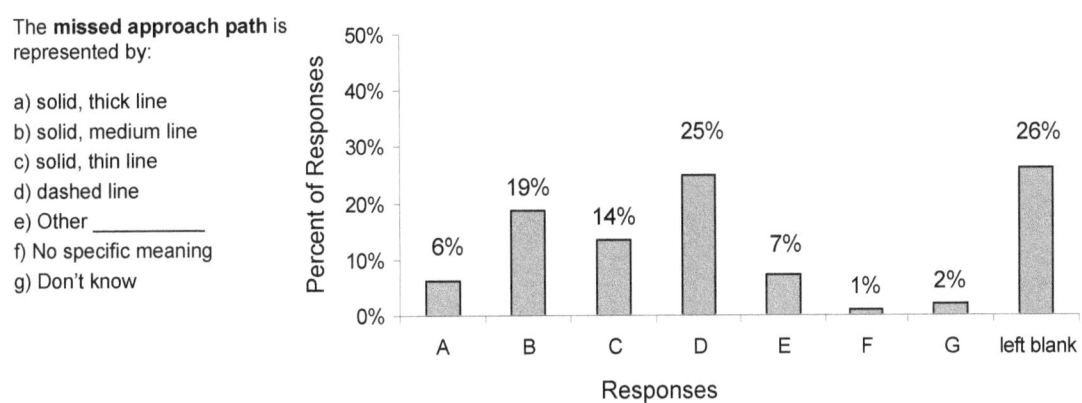

Figure 16. Responses for how a missed approach path is shown on electronic moving map displays.

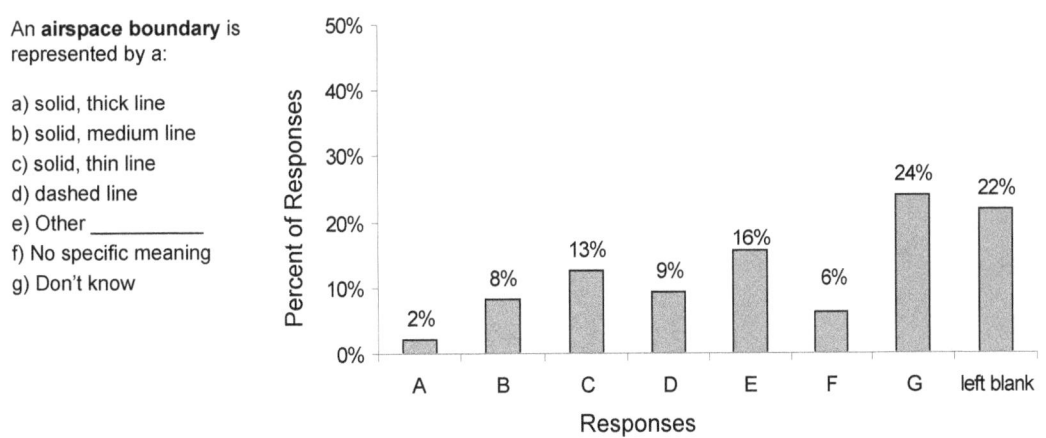

Figure 17. Responses for how airspace boundaries are shown on electronic moving map displays.

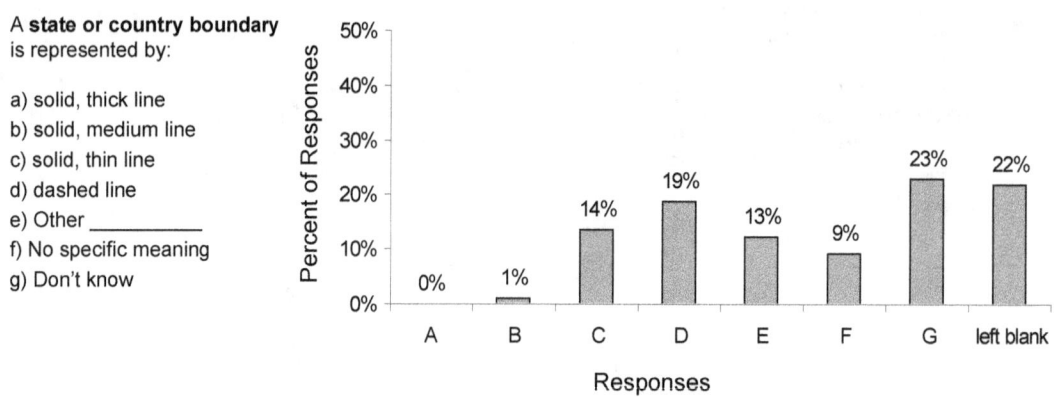

Figure 18. Responses for how a state or country boundary is shown on electronic moving map displays.

7 Summary and Conclusions

Four research questions related to development of recommendations for electronic aeronautical chart symbology (i.e., symbols and lines) were studied. Note that this research applies to any electronic display that shows the symbology tested in this study, regardless of the intended function of the display.

The main research question addressed pilot identification of several proposed symbols. Pilots generally identified the test symbols, but a few problematic symbols were identified. Factors contributing towards the correct and incorrect interpretation of the confusing symbols were explored. One such factor was the similarity of the tested symbol shape to other symbols in use. If the test shape was similar to a well-known symbol, the shape similarity appeared to be helpful in recognizing the tested symbol. For example, the fact that the generic localizer shapes were similar to the well-known front and back-course localizer shapes potentially increased their recognition. However, if the test symbol was similar in shape to another symbol that represented a substantively different element, the shape similarity could have increased misinterpretation. For example, the double-ringed NDB shape was perhaps *too* similar to the shape for the military airport symbol.

Pilot's familiarity with a symbol, in terms of both its shape and use, also contributed to symbol recognition. Note that familiarity of a symbol in the aeronautical environment is affected by frequency of exposure and use; symbols that are not often seen and used, such as the MSA, will be harder to recognize. Symbol labels, or contextual information, may also give the pilot valuable clues for identifying less familiar shapes. For the airport beacon shape, for example, pilots may have needed additional information, such as a text descriptor, for better recognition.

The second research question in this study was whether labels increase the accuracy of identifying navigation aids. For four of the five tested symbols, there was no difference in participants' accuracy of symbol identification with or without the labeling information. However, this test was not definitive. A more detailed study is necessary to understand the full effect of labeling information on both the speed and accuracy of symbol identification.

The third research question explored the relationships that pilots see between navigation aids. The goal here was to identify symbol families that could be represented by a single generic symbol. The most common feature used to group the navigation aids was what information was provided (e.g., distance, radial, or point). Pilots also considered other features, such as the amount of information, its utility, and technology.

In the last part of the study, pilot knowledge of line style conventions in paper charts and electronic displays was explored. Pilots were fairly knowledgeable about paper chart conventions, although they could not reliably distinguish between similar line widths and dash lengths. Responses to the questions about line conventions on electronic moving map displays were more varied, which could indicate either that pilots did not know the conventions, or that the conventions are not well established on these displays. A future effort is planned to document current symbology used on electronic displays.

The SAE G-10 Aeronautical Charting Committee, and its members from the FAA and ICAO, will consider the results of this study as they work together to develop an updated industry recommendations document (ARP 5289). Input from the FAA sponsors and the SAE G-10 Aeronautical Charting Committee will determine what research direction is pursued. The overall goal of this research program is provide the FAA with data and recommendations for the development of standards for electronic aeronautical chart symbology. As issues arise, data will be collected to help identify the best way to address them.

8 References

Chandra, D. C., Yeh M., Riley, V. and Mangold, S.J. (2003). *Human factors considerations in the design and evaluation of Electronic Flight Bags (EFBs), Version 2* Cambridge, MA: USDOT Volpe National Transportation Systems Center. (NTIS No. PB2004101967)

Federal Aviation Administration, (2003) Technical Standard Order C165. Electronic Map Display Equipment for Graphical Depiction of Aircraft Position.

Federal Aviation Administration (2007a). *Instrument procedures handbook* (FAA-H-8261-1A). Washington, DC: Author.

Federal Aviation Administration (2007b). *Federal Aviation Regulations/Aeronautical information manual* (ASA-07-FR-AM-8K). Newcastle, WA: Aviation Supplies & Academics.

International Civil Aviation Organization (ICAO) Annex 4 (July, 2001). Aeronautical Charts. Annex 4 to the Convention on International Civil Aviation.

National Aeronautical Charting Office (NACO) (2004). Aeronautical Chart User's Guide 6th Edition. Washington D.C.: Federal Aviation Administration.

RTCA (2003). Minimum Operational Performance Standards for the Depiction of Navigational Information on Electronic Maps. DO-257A. Washington, D.C., RTCA.

Society of Automotive Engineers (SAE) (1997). *Electronic Aeronautical Symbols*, ARP 5289. Warrendale, PA, SAE.

Yeh, M. and D. C. Chandra (2004). Issues in symbol design for electronic displays of navigation information. *Proceedings of the 23rd Digital Avionics Systems Conference.* Salt Lake City, UT.

Yeh, M. and D. C. Chandra (2005). *Designing and Evaluating Symbols for Electronic Displays of Navigation Information: Symbol Stereotypes and Symbol-Feature Rules.* (Report Nos. DOT/FAA/AR-05/48, DOT-VNTSC-FAA-05-16) Cambridge, MA, US DOT Volpe National Transportation Systems Center.

Yeh, M. and Chandra, D. C. (2006). Pilot stereotypes for navigation symbols on electronic displays. In Reuzeau, F., Corker, K. & Boy, G. (Eds.) *Proceedings of the International Conference on Human-Computer Interaction in Aeronautics.* (pp. 66-73). Toulouse, France: Cépaduès-Éditions.

Appendix A: Questionnaire

Informed Consent

I, _____, understand that this study, entitled "Flight Symbology" is being conducted by the Volpe National Transportation Systems Center, United States Department of Transportation, and is being directed by Dr. Divya Chandra. This research is funded by the Federal Aviation Administration, Human Factors Research and Engineering Division.

Purpose of Study. There are many types of electronic displays that show navigation information (e.g., VORs and Global Position System (GPS) waypoints) to help pilots determine the aircraft's position. There are no standards in widespread use that ensure the compatibility of the symbols across all the various display platforms. The purpose of this study is to understand whether certain symbol shapes are recognizable and whether features encoded in the symbol shapes to convey additional information about the symbol are understandable and usable. A couple of charts are provided at the end of the questionnaire for you to use as reference.

Study Procedures. This experiment addresses the design of symbols and features used on chart and map displays. You will be shown a set of test symbol shapes and asked to identify the symbol or identify the feature conveyed by the symbol shape. For each question, you will be asked to indicate your level of confidence in your answer. The study is estimated to take less than an hour to complete.

Discomfort and Risks. The risks involved in your participation are low and do not exceed those you would experience working on your home computer for about one hour.

Benefits to You. Participation provides an opportunity to aid in the development of recommendations for the design of air transport and general aviation displays.

Participant Responsibilities. Please notify Dr. Divya Chandra (617-494-3882) if you experience any discomfort during the study.

In the Event of an Injury, we urge that you report any immediate or delayed injuries resulting from the study to Dr. Divya Chandra (617-494-3882).

Assurances and Rights of the Participant. Your participation in this study is completely voluntary. Your participation is strictly confidential, and no individual names or identities will be recorded with any data or released in any reports. Only arbitrary numbers are used to identify pilots who provide data.

If you have any questions, please let us know. For further information about this study, please feel free to contact:

Divya Chandra or Michelle Yeh

US DOT Volpe Center

55 Broadway

Cambridge, MA 02142

617.494.3882 / 617.494.3459

chandra@volpe.dot.gov

yeh@volpe.dot.gov

Statement of Consent

I have read this consent document. I understand its contents, and I freely consent to participate in this study under the conditions described. I may have a copy of this consent form if I request same.

Research Participant: _____ Date: _____

This page left blank intentionally.

Background Questionnaire

Age _____

Gender   Male   Female

Flight Hours Total _____   Average (per month) _____

                                            Last month _____

Instrument Time Total _____   Average (per month) _____

                                            Last month _____

Which manufacturer provides the charts that you use most? How long?
_____

Do you use charts from other manufacturers regularly? Which? How long?
_____

Ratings and Certificates: Please check the ratings and certificates that you have.

Airline Transport _____   Instrument _____

Commercial _____   Single Engine _____

Rotorcraft _____   Multi Engine _____

Please list other ratings that you hold: **List the three most recent aircraft you have flown**. Please list them in order from most frequent to least frequent.
Type Rating

_____   _____

_____   _____

_____   _____

_____

Flight Experience: Please check the type(s) of flying that you do:

Private   IFR _____   VFR _____

Air transport _____

Business _____

Corporate _____

Military _____

Do you have experience with the following:

- Glass cockpit?   Yes   No
- FMS?   Yes   No
- moving map displays?   Yes   No

If yes, what are the specific moving map display(s) you are most familiar with?

A.3

This page left blank intentionally.

Symbol Questionnaire

Part 1. Symbol Recall

A. General

The purpose of this task is to determine whether symbols being proposed for use on paper and electronic terminal procedure charts (e.g., SIDS, STARs) are understandable. For each shape below, identify it and indicate your level of confidence in your response. Some of the symbols are unusual, so you may not recognize all the symbols. Write "?" if you do not know or can't tell.

1.	⋀⋀	Symbol Type (or ?): _____ 1 2 3 4 5 6 7 Low Medium High Confidence Confidence Confidence
2.	人	Symbol Type (or ?): _____ 1 2 3 4 5 6 7 Low Medium High Confidence Confidence Confidence
3.	⚹	Symbol Type (or ?): _____ 1 2 3 4 5 6 7 Low Medium High Confidence Confidence Confidence
4.	◇∷○∷◇	Symbol Type (or ?): _____ 1 2 3 4 5 6 7 Low Medium High Confidence Confidence Confidence
5.	(VOR compass: 6600' / 10,500' / 8100' / 8600', 090°/180°/270°/360°, OED VOR)	Symbol Type (or ?): _____ 1 2 3 4 5 6 7 Low Medium High Confidence Confidence Confidence
6.	◉	Symbol Type (or ?): _____ 1 2 3 4 5 6 7 Low Medium High Confidence Confidence Confidence

7.		Symbol Type (or ?): _____ 1 2 3 4 5 6 7 Low Medium High Confidence Confidence Confidence
8.		Symbol Type (or ?): _____ 1 2 3 4 5 6 7 Low Medium High Confidence Confidence Confidence
9.		Symbol Type (or ?): _____ 1 2 3 4 5 6 7 Low Medium High Confidence Confidence Confidence
10.		Symbol Type (or ?): _____ 1 2 3 4 5 6 7 Low Medium High Confidence Confidence Confidence
11.		Symbol Type (or ?): _____ 1 2 3 4 5 6 7 Low Medium High Confidence Confidence Confidence
12.		Symbol Type (or ?): _____ 1 2 3 4 5 6 7 Low Medium High Confidence Confidence Confidence

13.	(lens shape)	Symbol Type (or ?): _____ 1 2 3 4 5 6 7 Low Medium High Confidence Confidence Confidence
14.	(dotted circle)	Symbol Type (or ?): _____ 1 2 3 4 5 6 7 Low Medium High Confidence Confidence Confidence
15.	(360°/180° racetrack)	Symbol Type (or ?): _____ 1 2 3 4 5 6 7 Low Medium High Confidence Confidence Confidence
16.	(CRW 25 NM compass with 1500/2200/4500/2500)	Symbol Type (or ?): _____ 1 2 3 4 5 6 7 Low Medium High Confidence Confidence Confidence
17.	(half-circle with 2000/4200, 15 NM)	Symbol Type (or ?): _____ 1 2 3 4 5 6 7 Low Medium High Confidence Confidence Confidence
18.	★	Symbol Type (or ?): _____ 1 2 3 4 5 6 7 Low Medium High Confidence Confidence Confidence

B. Airport Identification

Each of the following shapes may be used to represent an airport. For each symbol, identify the *type* of airport the symbol represents and indicate your level of confidence in your response. Write "?" if you do not know or can't tell.

1.	◎	Airport Type (or Can't Tell): _____ 1 2 3 4 5 6 7 Low Medium High Confidence Confidence Confidence

A.7

2.	⊕	Airport Type (or Can't Tell): _____ 1 2 3 4 5 6 7 Low Medium High Confidence Confidence Confidence
3.	Ⓗ	Airport Type (or Can't Tell): _____ 1 2 3 4 5 6 7 Low Medium High Confidence Confidence Confidence
4.	⊚Ⓗ	Airport Type (or Can't Tell): _____ 1 2 3 4 5 6 7 Low Medium High Confidence Confidence Confidence
5.	⊖	Airport Type (or Can't Tell): _____ 1 2 3 4 5 6 7 Low Medium High Confidence Confidence Confidence
6.	⊕⚓	Airport Type (or Can't Tell): _____ 1 2 3 4 5 6 7 Low Medium High Confidence Confidence Confidence
7.	⊕◎	Airport Type (or Can't Tell): _____ 1 2 3 4 5 6 7 Low Medium High Confidence Confidence Confidence
8.	⊗	Airport Type (or Can't Tell): _____ 1 2 3 4 5 6 7 Low Medium High Confidence Confidence Confidence

Part 2. Navigation Aids

A. Symbol Alone

The purpose of this task is to determine whether symbols being proposed for navigation aids are understandable. Each of the symbol shapes below may be used to represent a navigation aid. For each shape, identify it and indicate your level of confidence in your response. Include as much detail as you can in your answer (e.g., try to distinguish between VORs and VORDMEs). Circle only one.

1.	⬠	a) DME only c) VOR only e) VORTAC b) TACAN only d) VORDME f) Can't Tell (?) -- 1 2 3 4 5 6 7 Low Medium High Confidence Confidence Confidence
2.	☐	a) DME only c) VOR only e) VORTAC b) TACAN only d) VORDME f) Can't Tell (?) -- 1 2 3 4 5 6 7 Low Medium High Confidence Confidence Confidence
3.	⬡	a) DME only c) VOR only e) VORTAC b) TACAN only d) VORDME f) Can't Tell (?) -- 1 2 3 4 5 6 7 Low Medium High Confidence Confidence Confidence
4.	⬠	a) DME only c) VOR only e) VORTAC b) TACAN only d) VORDME f) Can't Tell (?) -- 1 2 3 4 5 6 7 Low Medium High Confidence Confidence Confidence

5.

a) DME only   c) VOR only   e) VORTAC
b) TACAN only   d) VORDME   f) Can't Tell (?)

```
1    2    3    4    5    6    7
Low           Medium           High
Confidence    Confidence    Confidence
```

B. Symbol with Additional Information

Each row below presents information that represents a navigation aid. The information may be the one or more of the following: symbol shape, identifier, frequency, or channel.

Based on the information provided, provide your best guess as to what the navigation aid is and indicate your level of confidence in your response. The navigation aid may be a DME only, TACAN only, VOR only, VORDME, or VORTAC. Include as much detail as you can in your answer (e.g., try to distinguish between VORs and VORDMEs). Note: The identifiers, channels, and frequencies here are fictional. Any overlap with real navigation aids is coincidental.

1. 115.8 Chan 105

a) DME only   c) VOR only   e) VORTAC
b) TACAN only   d) VORDME   f) Can't Tell (?)

```
1    2    3    4    5    6    7
Low           Medium           High
Confidence    Confidence    Confidence
```

2. 110.2

a) DME only   c) VOR only   e) VORTAC
b) TACAN only   d) VORDME   f) Can't Tell (?)

```
1    2    3    4    5    6    7
Low           Medium           High
Confidence    Confidence    Confidence
```

3. 425

a) DME only   c) VOR only   e) VORTAC
b) TACAN only   d) VORDME   f) Can't Tell (?)

```
1    2    3    4    5    6    7
Low           Medium           High
Confidence    Confidence    Confidence
```

A.10

4.	WATE	a) DME only c) VOR only e) VORTAC b) TACAN only d) VORDME f) Can't Tell (?) -- 1 2 3 4 5 6 7 Low Medium High Confidence Confidence Confidence
5.	108.2 Chan 19	a) DME only c) VOR only e) VORTAC b) TACAN only d) VORDME f) Can't Tell (?) -- 1 2 3 4 5 6 7 Low Medium High Confidence Confidence Confidence
6.	113.8	a) DME only c) VOR only e) VORTAC b) TACAN only d) VORDME f) Can't Tell (?) -- 1 2 3 4 5 6 7 Low Medium High Confidence Confidence Confidence
7.	EPHI	a) DME only c) VOR only e) VORTAC b) TACAN only d) VORDME f) Can't Tell (?) -- 1 2 3 4 5 6 7 Low Medium High Confidence Confidence Confidence
8.	ABC	a) DME only c) VOR only e) VORTAC b) TACAN only d) VORDME f) Can't Tell (?) -- 1 2 3 4 5 6 7 Low Medium High Confidence Confidence Confidence

9.	Chan 31	a) DME only　　c) VOR only　　e) VORTAC b) TACAN only　d) VORDME　　f) Can't Tell (?) -- 　1　　　2　　　3　　　4　　　5　　　6　　　7 　Low　　　　　　　Medium　　　　　　　High Confidence　　　Confidence　　　Confidence
10.	⬡ CCD	a) DME only　　c) VOR only　　e) VORTAC b) TACAN only　d) VORDME　　f) Can't Tell (?) -- 　1　　　2　　　3　　　4　　　5　　　6　　　7 　Low　　　　　　　Medium　　　　　　　High Confidence　　　Confidence　　　Confidence
11.	OKT	a) DME only　　c) VOR only　　e) VORTAC b) TACAN only　d) VORDME　　f) Can't Tell (?) -- 　1　　　2　　　3　　　4　　　5　　　6　　　7 　Low　　　　　　　Medium　　　　　　　High Confidence　　　Confidence　　　Confidence
12.	⬠ KANGA	a) DME only　　c) VOR only　　e) VORTAC b) TACAN only　d) VORDME　　f) Can't Tell (?) -- 　1　　　2　　　3　　　4　　　5　　　6　　　7 　Low　　　　　　　Medium　　　　　　　High Confidence　　　Confidence　　　Confidence
13.	⬡ 112.6	a) DME only　　c) VOR only　　e) VORTAC b) TACAN only　d) VORDME　　f) Can't Tell (?) -- 　1　　　2　　　3　　　4　　　5　　　6　　　7 　Low　　　　　　　Medium　　　　　　　High Confidence　　　Confidence　　　Confidence

14.	☐ 105.0 GGH	a) DME only c) VOR only e) VORTAC b) TACAN only d) VORDME f) Can't Tell (?) -- 1 2 3 4 5 6 7 Low Medium High Confidence Confidence Confidence
15.	TRIMP Chan 45	a) DME only c) VOR only e) VORTAC b) TACAN only d) VORDME f) Can't Tell (?) -- 1 2 3 4 5 6 7 Low Medium High Confidence Confidence Confidence
16.	☐ Chan 26	a) DME only c) VOR only e) VORTAC b) TACAN only d) VORDME f) Can't Tell (?) -- 1 2 3 4 5 6 7 Low Medium High Confidence Confidence Confidence
17.	⬡ 365 OTI	a) DME only c) VOR only e) VORTAC b) TACAN only d) VORDME f) Can't Tell (?) -- 1 2 3 4 5 6 7 Low Medium High Confidence Confidence Confidence
18.	⬡ 109.5 OOM	a) DME only c) VOR only e) VORTAC b) TACAN only d) VORDME f) Can't Tell (?) -- 1 2 3 4 5 6 7 Low Medium High Confidence Confidence Confidence

A.13

19.	⬡ 112.4 Chan 71	a) DME only c) VOR only e) VORTAC b) TACAN only d) VORDME f) Can't Tell (?) --- 1 2 3 4 5 6 7 **Low** **Medium** **High** **Confidence** **Confidence** **Confidence**
20.	⬠ Chan 88	a) DME only c) VOR only e) VORTAC b) TACAN only d) VORDME f) Can't Tell (?) --- 1 2 3 4 5 6 7 **Low** **Medium** **High** **Confidence** **Confidence** **Confidence**
21.	⬠ ROBBNS Chan 90	a) DME only c) VOR only e) VORTAC b) TACAN only d) VORDME f) Can't Tell (?) --- 1 2 3 4 5 6 7 **Low** **Medium** **High** **Confidence** **Confidence** **Confidence**
22.	110.2 FES Chan 39	a) DME only c) VOR only e) VORTAC b) TACAN only d) VORDME f) Can't Tell (?) --- 1 2 3 4 5 6 7 **Low** **Medium** **High** **Confidence** **Confidence** **Confidence**
23.	▢ DAIZE Chan 15	a) DME only c) VOR only e) VORTAC b) TACAN only d) VORDME f) Can't Tell (?) --- 1 2 3 4 5 6 7 **Low** **Medium** **High** **Confidence** **Confidence** **Confidence**

24.	110.4 JBC	a) DME only c) VOR only e) VORTAC b) TACAN only d) VORDME f) Can't Tell (?) -- 1 2 3 4 5 6 7 Low Medium High Confidence Confidence Confidence
25.	113.0 ANT Chan 77	a) DME only c) VOR only e) VORTAC b) TACAN only d) VORDME f) Can't Tell (?) -- 1 2 3 4 5 6 7 Low Medium High Confidence Confidence Confidence
26.	FFG	a) DME only c) VOR only e) VORTAC b) TACAN only d) VORDME f) Can't Tell (?) -- 1 2 3 4 5 6 7 Low Medium High Confidence Confidence Confidence
27.	115.0 PQR Chan 97	a) DME only c) VOR only e) VORTAC b) TACAN only d) VORDME f) Can't Tell (?) -- 1 2 3 4 5 6 7 Low Medium High Confidence Confidence Confidence
28.	114.7 BCC Chan 94	a) DME only c) VOR only e) VORTAC b) TACAN only d) VORDME f) Can't Tell (?) -- 1 2 3 4 5 6 7 Low Medium High Confidence Confidence Confidence

Part 3. Symbol Classification

We are interested in the relationships that pilots see among navigation aids. Eight navigation aids are listed. Your task is to divide the symbol types below into groups, according to the instructions. Name each group to explain the rule you used to divide the symbols into groups.

1. Divide the symbols below into **2** groups.
   - a) DME
   - b) Fix
   - c) NDB
   - d) TACAN
   - e) VOR
   - f) VORDME
   - g) VORTAC
   - h) Waypoint

Group 1	Group 2
Name:	Name:

2. Divide the symbols below into **3** groups.
   - a) DME
   - b) Fix
   - c) NDB
   - d) TACAN
   - e) VOR
   - f) VORDME
   - g) VORTAC
   - h) Waypoint

Group 1	Group 2	Group 3
Name:	Name:	Name:

3. Divide the symbols below into **4** groups.
   a) DME
   b) Fix
   c) NDB
   d) TACAN
   e) VOR
   f) VORDME
   g) VORTAC
   h) Waypoint

Group 1	Group 2	Group 3	Group 4
Name:	Name:	Name:	Name:

2. Divide the symbols below into **5** groups.
   a) DME
   b) Fix
   c) NDB
   d) TACAN
   e) VOR
   f) VORDME
   g) VORTAC
   h) Waypoint

Group 1	Group 2	Group 3	Group 4	Group 5
Name:	Name:	Name:	Name:	Name:

Part 4. Line Styles

The purpose of this task is to determine whether line style conventions used on paper charts and moving map displays are understandable.

A. Paper Chart Styles

On a **paper chart**, solid thick, medium, and thin lines and dashed long, medium, and short lines are used to convey different paths. For each path described below, select the line style(s) that you think best describe how it appears **when it appears on the plan view of a paper chart**.

1.	**A radial or a bearing line that defines an intersection** is represented by a: a) solid, thick line b) solid, medium line c) solid, thin line d) dashed, long line e) dashed, medium line f) dashed, short line g) Other _____ h) Don't know
2.	**The procedure path** is represented by a: i) solid, thick line j) solid, medium line k) solid, thin line l) dashed, long line m) dashed, medium line n) dashed, short line o) Other _____ p) Don't know
3.	A **missed approach path** is represented by a: q) solid, thick line r) solid, medium line s) solid, thin line t) dashed, long line u) dashed, medium line v) dashed, short line w) Other _____ x) Don't know
4.	A **transition path** is represented by a: y) solid, thick line z) solid, medium line aa) solid, thin line bb) dashed, long line cc) dashed, medium line dd) dashed, short line ee) Other _____ ff) Don't know

5.	A **visual flight path** is represented by a:	
	gg)	solid, thick line
	hh)	solid, medium line
	ii)	solid, thin line
	jj)	dashed, long line
	kk)	dashed, medium line
	ll)	dashed, short line
	mm)	Other _____
	nn)	Don't know
6.	**Other conditional routes** are represented by a:	
	oo)	solid, thick line
	pp)	solid, medium line
	qq)	solid, thin line
	rr)	dashed, long line
	ss)	dashed, medium line
	tt)	dashed, short line
	uu)	Other _____
	vv)	Don't know
7.	**Airways** are represented by a:	
	ww)	solid, thick line
	xx)	solid, medium line
	yy)	solid, thin line
	zz)	dashed, long line
	aaa)	dashed, medium line
	bbb)	dashed, short line
	ccc)	Other _____
	ddd)	Don't know

B. Moving Map Display Styles (Skip this section if you are not familiar with moving map displays)

On a **moving map display**, solid thick, medium, and thin lines and dashed lines are used to convey different paths. For each path described below, select the line style(s) that you think best describe how it appears **when it appears on a moving map display**.

1.	The **missed approach path** is represented by a:	
	h)	solid, thick line
	i)	solid, medium line
	j)	solid, thin line
	k)	dashed line
	l)	Other _____
	m)	No specific meaning
	n)	Don't know

2.	**Airways** are represented by a:	
	a) solid, thick line	
	b) solid, medium line	
	c) solid, thin line	
	d) dashed line	
	e) Other _____	
	f) No specific meaning	
	g) Don't know	
3.	The **alternate (non-activated) flight plan** is represented by a:	
	a) solid, thick line	
	b) solid, medium line	
	c) solid, thin line	
	d) dashed line	
	e) Other _____	
	f) No specific meaning	
	g) Don't know	
4.	A **state or country boundary** is represented by a	
	a) solid, thick line	
	b) solid, medium line	
	c) solid, thin line	
	d) dashed line	
	e) Other _____	
	f) No specific meaning	
	g) Don't know	
5.	The **active flight plan** is represented by a:	
	a) solid, thick line	
	b) solid, medium line	
	c) solid, thin line	
	d) dashed line	
	e) Other _____	
	f) No specific meaning	
	g) Don't know	
6.	An **airspace boundary** is represented by a:	
	a) solid, thick line	
	b) solid, medium line	
	c) solid, thin line	
	d) dashed line	
	e) Other _____	
	f) No specific meaning	
	g) Don't know	

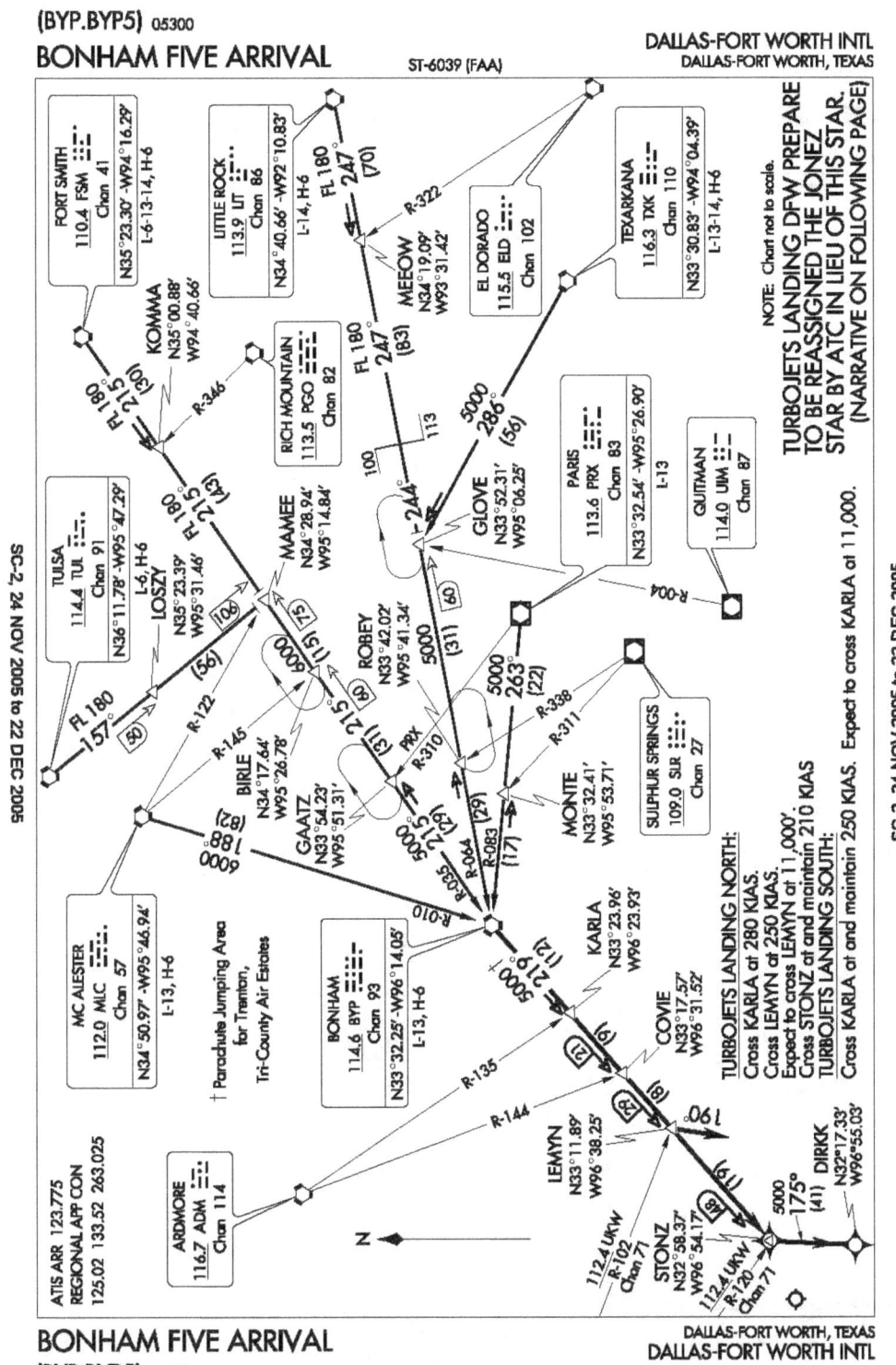

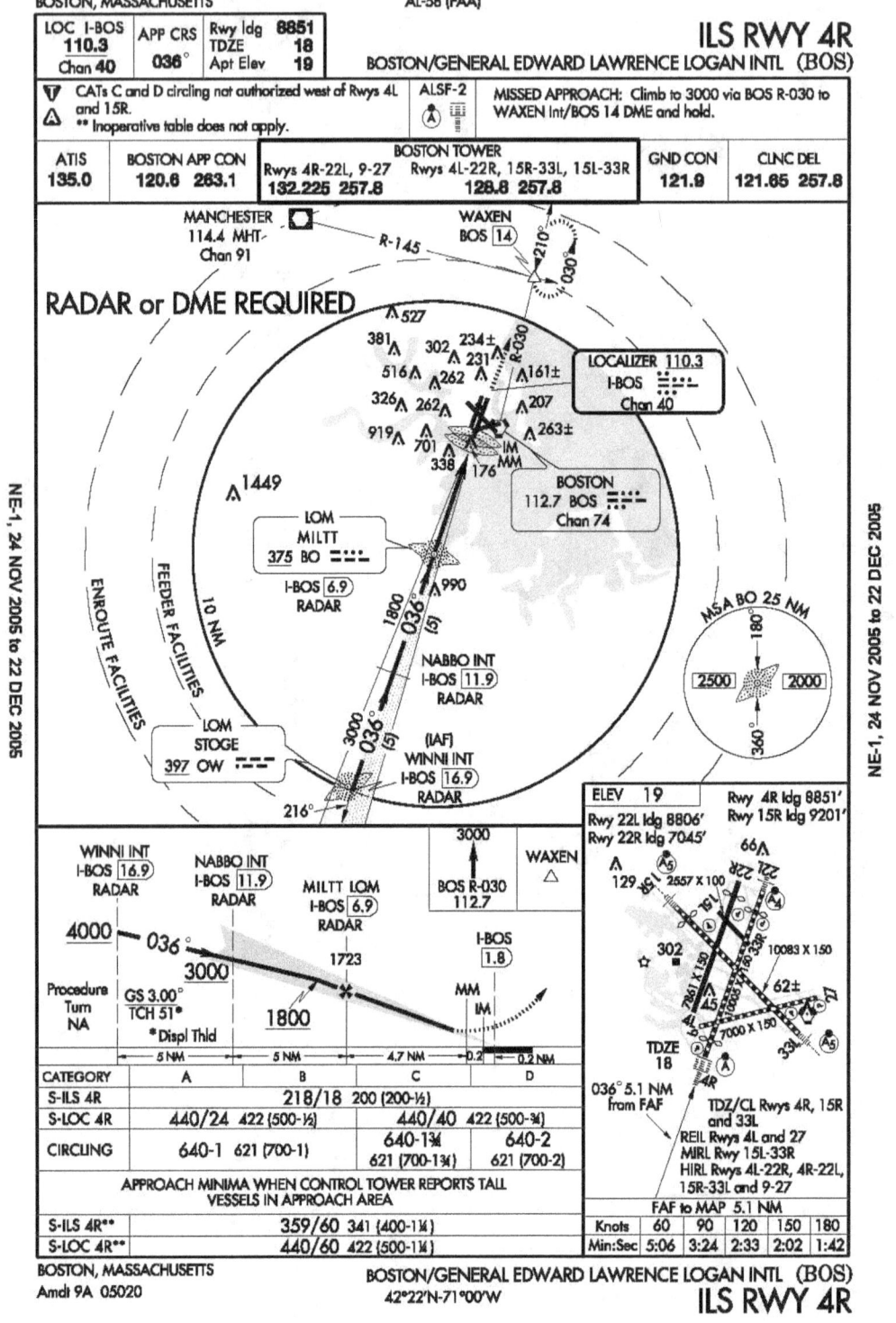

Appendix B:
Response Coding Method for the
General-Symbol Identification Task

The purpose of this appendix is to explain the process used for coding the raw response to the identification task for general symbols.

The raw responses to the symbol identification task produced multiple unique responses. Some of these were simple variants of each other (e.g., use of abbreviations) but some responses clearly represented a different concept from other responses. The first step in the response coding was to identify how many and which unique concepts were contained in all of the individual responses.

To do this, the individual responses were first sorted alphabetically, and duplicate responses were eliminated. The next step was to determine which unique responses belonged together in one category. The experimenters worked with the SAE G-10 Aeronautical Charting Committee to develop the categories and to identify what responses belonged in what category, particularly if there was some uncertainty. The final result was a set of response categories that included all of the individual responses. Note that there were typically a few unrelated unique responses that were coded together in a miscellaneous response category titled *Other*. A question-mark response was categorized as *Can't Tell*. If there was no response at all, the cell was coded as missing data.

Table 23 below is an excerpt from a raw data file for one of the symbols, the low single obstruction symbol (i.e., item #10 from the General Symbols section of the symbol identification task in Appendix A). For this symbol, the response categories used were:

A   Obstruction
B   Manmade Structures
C   Terrain
D   Other

As shown in Table 23, responses containing the word *obstacle* were placed in category A. Responses such as *multiple towers* and *antennae* were placed in category B. Responses relating to terrain were placed in the category C. These response category codes are shown under the column titled *General Answer*. Note that some responses were coded in more than one response category. For example the response from Participant #32 mentions both *single tower* (a manmade structure, category B), and *obstacle* response category A. Therefore that response was coded as "A, B." The response from Participant #51 was coded as "A, C" because it mentions both an obstacle and terrain.

For some symbols, the response was also coded in terms of whether details about the symbol were mentioned or not. For example, for the obstruction symbol, details about the mention of height and multiplicity were also coded. The General Answer for the response was coded without regard to the detail provided. For example, any response that said *obstacle* or *obstruction* counted as an obstacle (A), even if the detail was incorrect. The response from Participant #22 (in the first row of Table 23) provides a specific example. The response given was *obstacles* which was classified as the correct general answer (A) even though the detail was not correct because more than one obstacle was mentioned. Because the detail about multiplicity was incorrect, Participant #22 did not get an "x" under the detail column titled *Single*.

Similarly, the response *Obstruction > 1000 AGL* (from Participant #39) was classified as an obstacle (A) with correct mention of the *Single* detail because the word *obstruction*, not *obstructions*, was used. However, because the height detail reported was incorrect, the response was not counted towards the reports of correct height detail (in the column titled *Low*). The number of times incorrect detail was reported was not analyzed.

Item # Participant ID	Obstruction - Low Single 10	General Answer	Single	Low	Confidence
22	obstacles	A			5
23	Obstruction below 1000' AGL	A	x	x	6
27	Building	B	x		6
28	Localizer/Glidescope	D			4
29	Unidentified man made structure	B	x		6
30	obstacle/obstruction	A	x		4
31	Tower - Low Structure	B	x	x	6
32	single tower. obstacle	A, B	x		6
33	obstruction	A	x		
34	tower	B	x		4
35	obstruction - low height	A	x	x	5
36	Obstacle	A	x		4
37	obstacle	A	x		7
38	Single obstruction shorter than 1000' AGL	A	x	x	7
39	Obstruction > 1000' AGL	A	x		7
40	Twr	B	x		5
41	Tower	B	x		6
44	Terrain	C			7
45	unidentified man made structure	B	x		7
46	Tower <500'	B	x	x	7
47	Man made obstruction in 400'AGL with lights	B	x		
48	obstacle	A	x		6
49	Antennae	B			5
50	Tower	B	x		5
51	Obstacle on High Terrain	A, C	x		6
52	Single Obstruction	A	x		7
53	Single manmade obstruction	A, B	x		7

Table 23. Excerpt of data for low single obstruction symbol.

Appendix C:
Cluster Analysis Results

The tables below present the similarity matrices for the symbol classification task. The navigation aids are listed in the columns and rows of each matrix, and the number of times two navigation aids were grouped together is shown in the cells. The upper and lower halves of the matrices mirror each other, and cells along the diagonal indicate how often the navigation aid was placed in a group by itself.

Dendrograms from the cluster analysis that depict the relationship among navigation aids for that grouping are shown below each matrix. A comparison of the dendrograms to the similarity matrices show the relationships participants perceived among the navigation aids. An explanation of how to interpret the dendrogram is included for the classification of navigation aids into two groups. The other dendrograms would be interpreted similarly.

Results for Classification into Two Groups

	DME	Fix	NDB	TACAN	VOR	VORDME	VORTAC	Waypoint
DME	3	17	65	70	65	68	66	13
Fix	17	--	20	12	16	18	17	89
NDB	65	20	--	63	75	61	58	18
TACAN	70	12	63	--	74	73	76	12
VOR	65	16	75	74	--	68	65	16
VORDME	68	18	61	73	68	--	79	16
VORTAC	66	17	58	76	65	79	--	17
Waypoint	13	89	18	12	16	16	17	--

```
     C A S E        0         5        10        15        20        25
   Label     Num    +---------+---------+---------+---------+---------+

   fix         2    ─┐
   waypoint    8    ─┘                                                │
   NDB         3    ───────┐                                          │
   VOR         5    ───────┤                                          │
   VORDME      6    ─────┐ │                                          │
   VORTAC      7    ─────┤ │                                          │
   TACAN       4    ─────┘ │                                          │
   DME         1    ───────┘                                          │
```

The dendrogram for the classification of navigation aids into two groups shows two main "clusters": one containing the fix and waypoint, and the other consisting of the NDB, VOR, VORDME, VORTAC, TACAN, and DME. The distance between clusters (i.e., the similarity between the items is shown along the horizontal axis and highlights the "closeness" of the fix and waypoint. Overall, the greatest number of clusters possible was 5 at a distance of 1: (1) fix and waypoint, (2) NDB and VOR, (3) VORDME and VORTAC, (4) TACAN only, and (5) DME only. The distinctness of a cluster is shown by the distance along the horizontal axis from the point at which is comes into existence to the point at which it becomes part of a larger cluster. Thus, pilots' categorizations showed that they considered the NDB and VOR to be closely related, as were the VORDME and VORTAC. At approximately a distance of 5, the VORDME and VORTAC joined with the TACAN to form a cluster, and at a distance of 8, this cluster was joined by the DME.

Results for Classification into Three Groups

	DME	Fix	NDB	TACAN	VOR	VORDME	VORTAC	Waypoint
DME	9	11	29	38	39	26	24	8
Fix	11	1	5	6	3	8	7	77
NDB	29	5	11	14	56	17	14	6
TACAN	38	6	14	--	24	59	59	5
VOR	39	3	56	24	--	35	33	2
VORDME	26	8	17	59	35	--	65	8
VORTAC	24	7	14	59	33	65	--	8
Waypoint	8	77	6	5	2	8	8	1

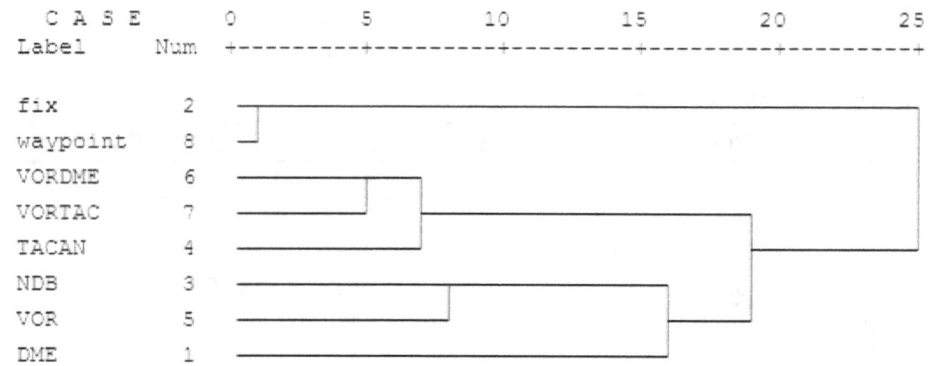

Results for Classification into Four Groups

	DME	Fix	NDB	TACAN	VOR	VORDME	VORTAC	Waypoint
DME	30	4	11	20	17	15	10	3
Fix	4	4	3	2	0	2	4	72
NDB	11	3	24	3	38	3	3	3
TACAN	20	2	3	12	8	39	44	1
VOR	17	0	38	8	--	27	22	0
VORDME	15	2	3	39	27	--	70	1
VORTAC	10	4	3	44	22	70	--	2
Waypoint	3	72	3	1	0	1	2	6

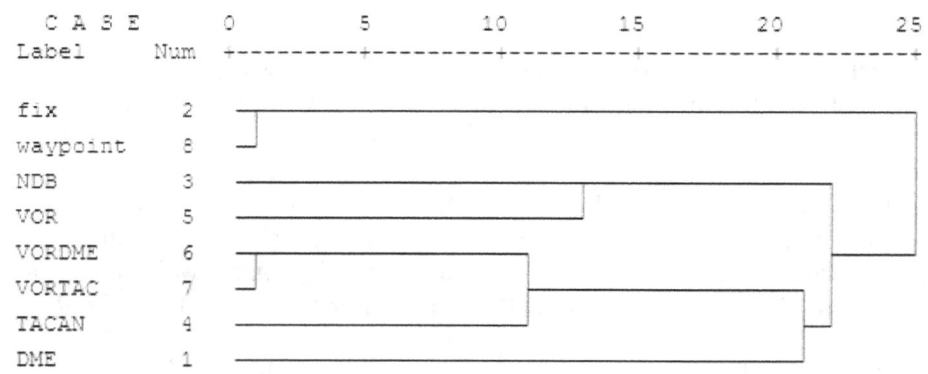

Results for Classification into Five Groups

	DME	Fix	NDB	TACAN	VOR	VORDME	VORTAC	Waypoint
DME	41	1	6	16	8	4	2	1
Fix	1	12	0	1	1	3	3	57
NDB	6	0	45	1	22	1	0	0
TACAN	16	1	1	24	5	21	26	1
VOR	8	1	22	5	26	13	11	0
VORDME	4	3	1	21	13	8	55	1
VORTAC	2	3	0	26	11	55	5	1
Waypoint	1	57	0	1	0	1	1	15

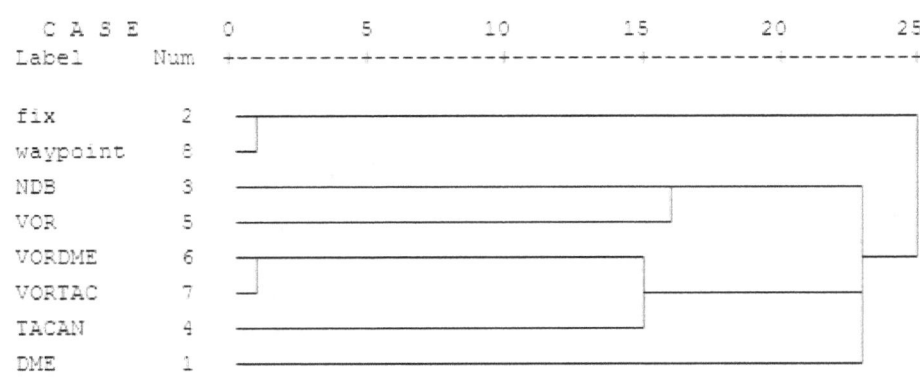

```
       C A S E      0         5        10        15        20        25
    Label    Num    +---------+---------+---------+---------+---------+

    fix       2     ─┐
    waypoint  8     ─┘                                                │
    NDB       3     ──────────────────────────────┐                   │
    VOR       5     ──────────────────────────────┤                   │
    VORDME    6     ─┐                            │                   │
    VORTAC    7     ─┤                            │                   │
    TACAN     4     ─┴────────────────────────────┘                   │
    DME       1     ──────────────────────────────────────────────────┘
```

www.ingramcontent.com/pod-product-compliance
Lightning Source LLC
Chambersburg PA
CBHW081847170526
45167CB00007B/2924